# IT
# SOUNDS
# LIKE FUN

## HOW TO USE AND ENJOY YOUR
## TAPE RECORDER AND STEREO

HERE'S ALL the information you need to select and begin using a tape recorder and stereo equipment. This book tells all about recorders, from cassettes to tape decks, and covers as well microphones, receivers, and loudspeakers. It tells you clearly how to record your friends speaking, singing, playing music; describes games and stunts to play with tape recorders; tells how to set up a formal interview; and how to use videotape equipment. This informative book will help you to get all the fun possible out of using your stereo and recording equipment.

# IT SOUNDS LIKE FUN

## HOW TO USE AND ENJOY YOUR TAPE RECORDER AND STEREO

*by Edward F. Dolan, Jr.*

Illustrations

JULIAN MESSNER
New York

Published by Julian Messner, a Simon & Schuster
Division of Gulf & Western Corporation,
Simon & Schuster Building,
1230 Avenue of the Americas,
New York, New York  10020.
JULIAN MESSNER and colophon are trademarks of
Simon & Schuster, registered in the U.S. Patent
and Trademark Office.

Manufactured in the United States of America.

Design by Irving Perkins Assoc.

Library of Congress Cataloging in Publication Data

Dolan, Edward F   1924–
It sounds like fun.

SUMMARY: Information for selecting and using tape
recorders and stereo equipment.
1. Magnetic recorders and recording—Juvenile
literature.   2. Stereophonic sound systems—Juvenile
literature.   [1. Tape recorders and recording.
2. Stereophonic sound systems]   I. Title.
TK9967.D64        621.389'.3        81-296

ISBN  0-671-34053-0        AACR1

# CONTENTS

# ACKNOWLEDGMENTS

I AM indebted to many people for their help in the preparation of this book. In particular, my thanks must go to Timothy L. Dolan and Richard B. Lyttle for their research assistance and editorial comment. They were most generous with the time that they gave to the book.

Special thanks are due to the many companies that provided me with information and photographs. They are:

Akai America, Ltd.; Ampex Corporation; Avid Corporation; Bang & Olufsen of America, Inc.; B.I.C/Avnet, a Division of Avnet, Inc.; Epicure Products, Inc.; Fisher Corporation; Garrard U.S.A.; General Electric Company; Heath Schlumberger, Heath Company; Hitachi Sales Corporation of America; Magnavox Consumer Electronics Company; Otari Corporation.

Panasonic Company; Philips Audio Video Systems Corporation; Philips High Fidelity Laboratories, Ltd.; Pioneer, U.S. Pioneer Electronics Corporation; Radio Shack, a Tandy Corporation; Sankyo Seiki (America) Inc.; Sansui Electronics Corporation; Tanberg of America, Inc.; TEAC Corporation of America; United Audio Products, Inc.; Yamaha International Corporation; and Zenith Radio Corporation.

For Ruth and George Walker
Good friends

# IT
## SOUNDS
# LIKE FUN

CHAPTER

1

# GETTING ACQUAINTED

YOU'VE BEEN bitten. There's almost no doubt about it.
Of all the books in the store or on the library shelves,
you picked this one to look at. That usually means just one
thing. Somewhere along the line, the recording bug has
nipped you, or—take warning—is about to.

But don't worry. Millions of people (of all ages, shapes,
and sizes) have been nipped by the same bug since that day
during World War II when German scientists discovered
how to preserve sound on a narrow strip of magnetic tape.
The victims haven't suffered a bit. They've had a great time
with tape recording. For some, it's been an enjoyable pas-
time; for others, an engrossing hobby. Very soon you'll be
doing all of the things they can do.

Perhaps you'll want to build a collection of tapes by
recording the music played on your favorite radio station.

13

82-42

Perhaps you'll want to transfer your best phonograph records to tape so that the discs won't wear out or begin to sound scratchy from too much playing. Or perhaps you'd like to do what a cousin of mine did. He taped a series of radio commercials. Then he cut the tape and spliced the parts of the commercials together in various ways. The result: some hilarious "ads" of his own. They were a big hit among his friends.

One of his masterpieces went like this (the slashes indicate the places where he spliced the tape together):

> Try the new puppy food /
> That's recommended by doctors everywhere /
> For getting thirty miles to the gallon. /
> It looks great, tastes great /
> And kills all bugs with a single spray. /
> Look for it on your grocer's shelf /
> All wrapped in /
> A brand new set of radial tires.

If you don't care to produce your own commercials, then perhaps you might wish to record your friends' voices at a party. Or interview a prominent person for your school newspaper or one of your classes. Or put your school band, debating team, or drama club on tape to show them how they sound. Or . . .

Well, the list is almost endless. And the chances are that you already have some ideas of your own.

Manufacturers of recording equipment like to say that anyone can get in on all this fun with just a push of the button that sets the tape moving through a recording machine. That's true. Today's machines are quite easy to operate. But the most fun is to be had if you really know what you're doing when you push the button and if you learn all that you can about tapes and their machines. Then you'll be sure of making excellent recordings.

And that's the purpose of this book: to start you on the way to making the best recordings possible. As is true of everything in this world, the best place to start is right at the beginning, with some basic information. So let's look first at the kinds of tape machines that you'll find in use today.

## TAPE MACHINES

Basically, there are two kinds of tape machines: the tape recorder and the tape deck. Except for one difference, you'll find that they're exactly alike, but that one difference is a big one.

The tape recorder contains all the equipment that you need to make a recording and then play it back so that you can listen to it. For playing back the recording—or *playback*, as it's most often called—the recorder has a built-in amplifier to strengthen the sound, plus a built-in or attached loudspeaker system to broadcast the sound.

The tape deck houses all the equipment necessary for making a recording. But—and here's the big difference—it doesn't carry any gear for playback. It doesn't contain enough of an amplification system for the job, nor does it have any loudspeakers of its own. To hear what you've recorded, you must tie the deck into an outside amplification and loudspeaker system. One of the easiest ways to do this is to connect the deck to a stereo set.

Once you understand the difference between a recorder and a deck, you'll find that both can be divided into other types. These types have to do with the ways that the tape runs through the machines. In all, there are three types: the open-reel unit, the cassette, and the 8-track cartridge unit.

The open-reel unit is the oldest of all recorders and decks. It employs two plastic or metal reels. The tape starts on one reel and passes to the second reel during recording. The first reel is called the supply reel, and its companion is the take-up reel. They're known as open reels because, quite simply, the tape on them is "open" to your touch.

Cassette recorders and decks also use two reels, with

**OPEN-REEL UNIT** (*Photo courtesy Radio Shack*)

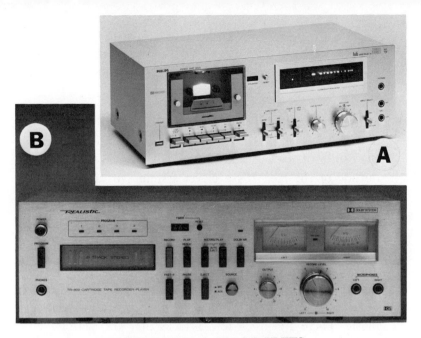

## CASSETTE AND 8-TRACK UNITS

A cassette deck (Picture A) is markedly similar in appearance to an 8-track unit (Picture B). Both are approximately the same size and both have many of the same controls. The most noticeable difference between the two is the housing in which the tape case is placed. (*Photo A courtesy Philips Audio Video Systems Corporation; Photo B courtesy Radio Shack*)

the tape traveling from one to the other during recording. But the reels are enclosed in a cassette, which is nothing more than a small, plastic case. Unlike the open reels, they and their tape can't be touched directly—not, that is, unless you go to the trouble of prying the little case open.

The 8-track cartridge unit was originally designed for people who wanted to listen to stereo music on commercially recorded tapes while in their cars. The name comes from the special tape that must always be used. The tape has eight channels, or tracks, running side by side all along its length. Recorded material can be placed in each of these channels.

The 8-track tape comes in a plastic case, called a car-

17

tridge, that is slightly larger than a cassette, but there is only one reel inside the cartridge. The tape unwinds from the center of the reel and then winds itself back onto the outside of the reel. It's a clever system (in fact, it looks downright miraculous when you see it at work) that allows the tape to be played again and again without ever stopping. For as long as the unit is playing, the tape keeps winding and rewinding itself.

Though the 8-track unit was first developed for use in cars, it is now found in many home stereo sets. It continues to be used mostly for listening to commercially recorded music. There are just a few 8-tracks that also permit you to make recordings of your own.

In the next two chapters, we'll take a closer look at the open-reel, cassette, and 8-track units to see exactly how they're constructed and how they work. But now that you've met them just briefly, let's answer a question that everyone new to recording always asks: "How on earth do any of these machines manage the miracle of putting sound on a piece of plastic ribbon?"

## FROM SOUND TO TAPE

The answer can begin with your voice. When you speak or sing or even grunt, the sounds that come out travel through the air in waves. Every sound creates these waves, and they're very much like the waves in an ocean. First, they have height. The height of each is determined by the loudness, or intensity, of the sound. Loud sounds produce high waves. Soft sounds produce low waves.

Next, like those in the ocean, the waves repeat themselves, doing so a number of times per second. These re-
peats are vibrations and are known as *cycles per second*
(cps), with the exact number in any second being determined by the bass or treble pitch of a sound. If you pitch
your voice down low, you'll produce just a few cps, perhaps
only thirty or forty. But, should you hit a high soprano note
or let loose with a scream, the wave might well cycle several
thousand times in a second.

Now let's see what happens to your sound waves when
you speak into a microphone that's connected to a recorder. Inside the microphone is a membrane device called
a diaphragm (more about this in Chapter Five). This
diaphragm begins to vibrate as soon as the sound waves hit
it. The vibrations cause electrical waves to take shape in
the current flowing through the wires leading from the
microphone to the recorder.

Formed by surges in voltage and fluctuations in the
current, these electrical waves are carbon copies of the
sound waves. They have the same height as the sound
waves, and the same number of cycles per second. Your
loud sounds become high-peaked electrical waves, and your
sounds that cycle forty times in a second become electrical
waves that cycle forty times in that same second. Literally,
your voice is transformed into electricity.

Both the sound waves and the electrical waves are also
known as *signals*, or *audio signals*. The second name comes
from the fact that the signals are meant to be heard.

In addition, the cycles made by an electrical wave have
a special name. They're called *Hertz*, in honor of Heinrich
Hertz, the nineteenth century German scientist famous for

his experiments with radio waves. The word is abbreviated to *Hz* and means a single cycle; 40 Hz, then, stands for 40 cycles per second, and 10,000 Hz means 10,000 cps. You'll also hear the terms *kilohertz* (kHz) and *megahertz* (MHz). The first stands for 1,000 cps, and the second for 1,000,000 cps.

On entering the recorder, the electrical waves are amplified. In a split second, they reach a device that's known as the recording head, or the tape head. With its two names, it stands at the heart of the recorder. In fact, it *is* the heart. We need to look at it for a moment before going on.

The head is actually an electromagnet. Built of iron (or some other material that can be easily magnetized) and then wrapped all around with a wire coil, it is usually shaped into a horseshoe or a circle. In either case, it is never fully closed, with a narrow space always being left between its

**RECORDING PROCESS**

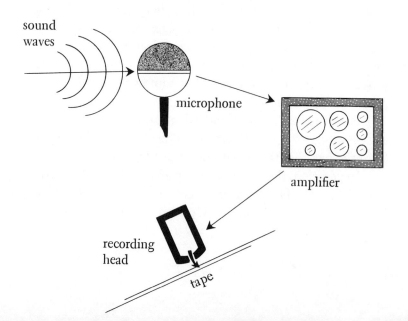

two ends. In today's recorders, that space usually measures no more than one-quarter of a millimeter—or .00025 of an inch—in width. This separation is the *recording gap*. The tape that comes traveling through the recorder brushes lightly against it.

Now back we go to the electrical waves. On reaching the tape head, they cause it to become magnetized. Instantly, as happens in any magnet, its ends become the north and south poles. Lines of magnetism flow through the head, escape from the north pole, and then try to get back in at the south pole.

In its own way, the flow of the magnetic lines is a carbon copy of the electrical waves. Though not wavelike, it imitates their height by altering its strength. When the waves are high-peaked, the flow is very strong. When they're low, it's weak.

At the same time, it imitates the cyclings per second. With every wave cycle, the north pole in the recording head changes places with the south pole and then jumps back again. And so, there in the gap, the magnetic lines flow back and forth in search of the south pole that will let them reenter the head. Their to-and-fro flow is in perfect rhythm with the cycling of the waves.

While all this is going on, the tape is brushing past the gap. Magnetic lines always find it difficult to travel through the air, even the tiny bit that's in the gap. To make things easier for themselves, they board the tape for the trip in search of the south pole. But it so happens that the tape is coated with particles of a substance that becomes magnetized when the lines touch it. Just as iron filings will arrange themselves into a pattern when placed near a magnet, so do the particles. They immediately form into patterns that correspond to the changing strengths of the magnetic flow and to the to-and-fro movements of the lines in the gap.

These patterns are invisible to the naked eye, but they've been photographed by high-powered cameras. They're seen to be clustered into barlike shapes along the tape.

In the instant that the patterns are formed, the recording process is complete. It's taken three steps to complete the job: your sound waves have been turned into matching electrical signals; the electrical signals have been turned into a matching flow of magnetism; and the flow has finished things off by leaving a matching pattern of magnetism on the tape. Your voice—transformed from sound to magnetism—is "locked" inside that pattern.

Suppose that you now turn the recorder off. The magnetism in the tape head immediately dies. This happens because the head is made of a material that is magnetically soft and unable to hold its magnetism without the help of an electric current. But the material on the tape has been chosen because it is magnetically hard, and so the magnetic pattern remains in place. The pattern will remain in place until you decide to get rid of it by running the tape past another head in the recorder—a special one called the erase head. The erase head hits the tape with an especially strong current that breaks the pattern up.

As can be seen in the next illustration, the erase head is always placed directly ahead of the tape head in a recording machine. This enables the erase head to eliminate all the previously recorded sounds while you're putting a new recording on the tape.

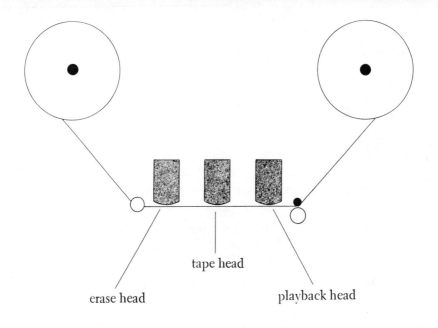

tape head

erase head     playback head

## ERASE, TAPE, AND PLAYBACK HEADS

But let's say that you don't wish to erase the tape just yet. Instead, you want to play it back to hear how you sound. You rewind the tape and send it through the recorder again. This time it brushes against what is known as the playback head. The reverse of the recording procedure now takes place. The magnetic pattern induces a small electrical current in the playback head. That electrical current is amplified and sent to a loudspeaker system. It emerges in sound waves from the loudspeakers, and there you are, listening to the sound of your own voice.

In some machines, as shown in the illustration, the tape and playback heads are separate units. But, in many other machines, you'll find them placed together in the same housing.

The recording and playback process described here works for all types of sound, whether it comes from the voice, a single musical instrument, an entire symphony orchestra, or from two cars crashing together out in the

23

street. And the process is used not just for the recording of sounds. It can just as easily put visual images and computer data on strips of tape.

One of the most popular tools in television is the video-tape recorder, which we'll be meeting in a later chapter. In a studio, a TV camera focuses on several actors. It picks up the light waves emanating from them, and the light waves are immediately turned into electrical signals; then, in their turn, the electrical signals are converted into magnetic impressions on the tape. From the tape, the images are flashed to your television screen at home.

Much the same kind of activity takes place when computer tapes are made. Electrical impulses that correspond to the symbols of the computer's code—its "language"—are transformed into magnetic impulses that implant themselves on the tape. On playback, the magnetic impulses are switched back into electrical impulses, with the symbols of the computer's code then being sent to a screen or a sheet of paper. There, they appear in our language. They may also appear as diagrams on the screen.

## MAGNETIC TAPE

Now that we've talked about how recorders and decks work the magic of putting sound on tape, we need to take time for another introduction. Let's meet the tape itself. It's known formally as magnetic tape.

The first tapes were made of paper, but it proved to be an unsatisfactory material, as everyone knew it would, because it could be so easily torn. Today's tapes are manufactured of plastic. In general, two kinds of plastic are most often used: acetate and polyester. You may hear some people talk also of mylar tapes, but mylar is really just another type of polyester.

Only one side of the tape is used for recording. Readily identified by its shiny look, it is coated with an adhesive solution containing a magnetic material. Though any magnetic substance may be used, only a few are able to give truly good recordings. The materials used today include ferric (iron) oxide, chromium dioxide, cobalt oxide, and ferri-chrome (a combination of iron oxide and chromium dioxide). Each substance offers its own advantages to the recordist, and in time you'll be able to pick the one that pleases you most.

No matter the substance used, it is ground into a fine powder by the manufacturer, after which he paints it as smoothly as possible onto the plastic ribbon. A smooth, uniform coating is necessary if the tape is to produce a good recording. Suppose, for instance, that the manufacturer leaves a thin spot or a blank spot in the coating; it will cause a *dropout*—the fading or, worse, the loss of a musical note or a syllable in a word. This type of mishap, though seemingly inconsequential, can ruin things for someone who wants the recording to be perfect. And that's what every serious recordist wants—perfection.

In addition to the substances already mentioned, tapes have recently been coated with pure metal particles. These tapes, though they produce excellent recordings, are not yet in wide use. They're fairly expensive and, for best results, should be played on machines that are specially built to handle them.

Also coming into use is digital tape. It is tape on which the sounds are implanted by electrical impulses in much the same way that information is placed on a computer tape. The digital tape system is called Pulse Code Modulation (PCM) and is presently being tried by a number of FM radio stations. Hardly anyone, however, expects that PCM will be generally available for use in home recordings until late in the 1980s.

All tapes are divided into tracks, or channels. These are narrow bands that run along the full length of the tape. They're the parts of the tape on which your recorded voice or music is implanted.

When the first recorders were manufactured years ago, the depth of their recording gaps extended all the way across the width of the tape, as can be seen in Figure A in the next illustration. The tape, then, could have just one track and was called a full-track or single-channel tape. On it were made what are known as monophonic—a technical term for single-channel—recordings.

Monophonic recordings were fine for a start, but they posed a problem that called for a change as soon as possible. The tape, with its full width being used for the recording, could travel through the machine just once, in one direction, before it was completely used up. It took no more than a few minutes to make the trip, meaning that either a great many tapes or some giant reels would be necessary for a recording session of any length. Either way, the expense of buying all that tape promised to be beyond the means of recordists who had to watch their pennies. If tape recording was to become widely popular, a method had to be found for putting more recorded material on an average size reel.

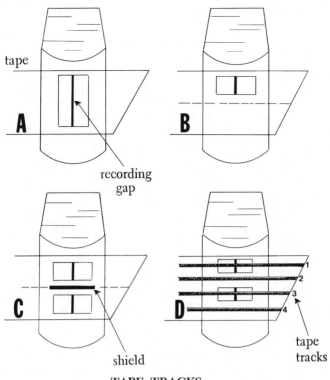

tape

A

B

recording
gap

C

D

shield

tape
tracks

**TAPE TRACKS**

In time, sound engineers solved the problem. They divided the tape into two channels all along its length (Figure B) and cut the recording gap down to less than half of its original depth. Christened as 2-track tape, the plastic ribbon now accepted a recording in one channel as it ran past the gap. At the end of the run, the tape was turned upside down by having the reels change places. Then back through the machine it went, with the recording being made this time in the second channel. A single tape could now handle twice as much recorded material as before.

Though each channel was used for a monophonic recording at first, the 2-track system opened the way for stereophonic tapes (Figure C). Sound engineers devised a recording head with two gaps, thus enabling both channels to be recorded on at the same time. To keep the magnetic patterns on one channel from flowing over and mixing with those on the other channel, a strip of shielding material was run all along the middle of the tape.

Let's pause a moment here to see how the stereo process works. Suppose that you're recording a performance by your school band. As the tape runs past the gaps, some of the sounds are imprinted on one channel while the others are sent to the second channel. You then play back both channels simultaneously. On coming out of loudspeakers that are set a proper distance apart, the music from the two channels surrounds you and blends into a sound that is particularly full and very realistic. It gives you a feeling that the monophonic recording can't give—the feeling that you're listening to the band in person.

Actually, no recorded music ever sounds exactly like live music. Some degree of reality is always lost. But expertly recorded stereophonic music comes as close to the real thing as we've been able to get so far. The realism of recorded sound is called high fidelity. The greater the realism of the sound, the higher the fidelity is said to be.

The first stereophonic tapes were marvelous to listen to, but they brought back the problem of old. Once again, now that both channels were being used at the same time, the tape could run through the recording machine in one direction only. Back to the drawing board the engineers went, this time to put more stereo material on each tape.

The answer was 4-track stereo (Figure D). As the name indicates, the tape was divided into four channels, and they were numbered 1, 2, 3, and 4. When the tape now ran through the machine in one direction, the gaps picked up two channels. The two remaining channels were picked up on the second run-through.

In some recording units today, the upper recording gap picks up channel 1 on the first run-through, while the lower gap takes care of channel 3. When the tape is turned over and sent through for the second time, the upper gap handles channel 4, with channel 2 traveling past the lower gap. In other units, the gaps pick up channels 1 and 2 on the first run, and then 3 and 4 when the tape is turned over.

Once the 4-track tape was developed, sound engineers realized that it would be possible to record all the channels simultaneously and then play them back through four speakers—with each speaker broadcasting one of the tracks —for a fuller sound. This system, which was popular a few years ago but which is not widely used today, is called 4-track quadraphonic sound.

The latest development is the 8-track tape. It employs a rather unusual head. We'll be talking about both the tape and the head when we get to the workings of the 8-track player in a later chapter.


Well, that does it for our introduction to tape recording. We've met the various kinds of recording machines and the tapes that travel through them, and we've talked about the scientific principles that enable them to perform their miracles. Now it's time to open up each of the machines and see how each makes its reels go around.

CHAPTER

# OPEN-REEL TALK

As you know, the open-reel machine is the oldest of all decks and recorders. Also, because of the fine engineering developments that have gone into it through the years, it is the most expensive unit on the market. Open-reel prices start at around $500 and run to $2,000 or more. For your money, you get a recording instrument that produces top-quality sounds, sounds that are superior to those coming from the cassette and the 8-track.

The purpose of this chapter is to find out how the open-reel unit works. To do this, we're going to look at its mechanical system, its many controls, and its tape.

## THE MECHANICAL SYSTEM

Every open-reel, cassette, and 8-track unit has both a mechanical and an electrical system. When we talked in Chapter One of how sounds are sent to the tape, we were actually talking about the electrical system. When we now talk about the mechanical system, we'll be talking of how the machine moves the tape past the recording gap.

The job of the mechanical system is not only to move the tape past the recording gap but also to move it past at a steady rate of speed. That steady rate is absolutely necessary if you're to have any hope of making a good recording. Should the transport system start to act up and cause the tape to dance along at varying speeds, you're going to hear some pretty odd things on playback. Sounds will accelerate at times in imitation of Walt Disney's cartoon characters or stretch out in wailing tones. These changes are known as *flutter* and *wow*. Flutter is caused by fast variations in speed, and wow by slow ones.

The mechanical system in an open-reel unit consists of four basic parts: a guide roller, a capstan, a pinch roller, and one or more electrical motors. Let's see how they work.

### OPEN-REEL TRANSPORT SYSTEM

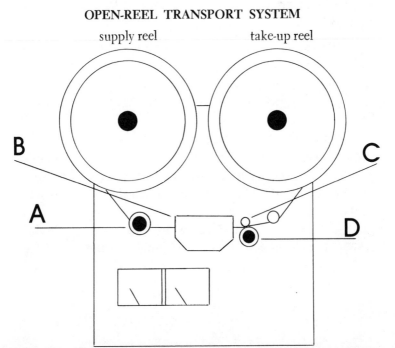

supply reel          take-up reel

B

C

A          D

The fun of recording begins when you thread the tape through the machine. As seen in the illustration, you run the tape from the supply reel to the guide roller (Figure A), then past the tape head (B), and then between the capstan (C) and the pinch roller (D). From there, it travels past another roller and rises to the take-up reel.

Obviously, the guide roller's job is to send the tape in a straight line to the recording head. It is simply a little spool that spins freely, with its edges upturned to keep the tape from sliding off. It may be made of metal, plastic, or hard rubber.

The pinch roller is also a little spool, but, because of the job we'll soon see it doing, it is fashioned of soft rubber, a rubber sometimes so soft that it seems spongelike. The capstan, on the other hand, is a metal shaft, circular in shape and made of fine quality steel. It is connected to a flywheel behind the face of the unit. The flywheel, in its turn, is connected to an electric motor by means of a belt and pulleys.

At the moment, prior to recording, the capstan and the pinch roller are separated by a narrow space. Running through the space, the tape rests lightly against the capstan. But now let's switch the unit on. Instantly, three things happen.

Behind the face panel, the electric motor springs to life and begins to turn the take-up reel slowly. The motor also activates the flywheel and causes the capstan to spin rapidly at a steady rate. And the pinch roller, with its soft rubber protecting the tape from damage, swings over and presses the tape against the capstan. The result of all this activity is that the tape is pulled steadily past the head. The pulling continues until the unit is switched off. The pinch roller then swings away from the capstan and leaves the little space once again between them.

The electric motor turns the take-up reel and the capstan during both recording and playback. It also performs a variety of other tasks. At the flick of a switch, it will run faster or slower so that the tape can be moved along at any of several recording speeds (more about these speeds in another few pages). At the flick of another switch, it will cause the supply reel to turn swiftly and rewind the tape back to its starting point. And still another switch will spin the take-up reel swiftly so that you can quickly advance the tape to some point that you want to use for recording or playback.

The illustration below shows the various belts that enable the motor to turn the reels and the capstan. While not in use, the belts are held in a slack position. At the moment when a reel or the capstan is to be turned, a proper amount of tension is automatically applied to the appropriate belt.

### OPEN-REEL MOTOR SYSTEM

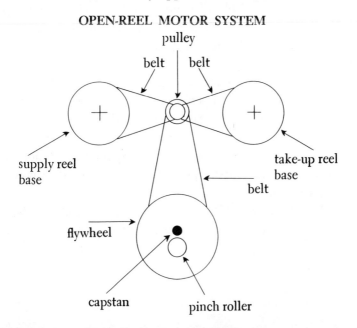

Shown in the illustration is a unit operated by just one motor. Many of today's models employ two or three motors. When two motors are used, one takes care of the capstan, leaving the reels to the other. Three-motor units assign one motor each to the capstan, the supply reel, and the take-up reel.

In some of the more expensive machines, the motors do not turn the capstan and reels by means of belts. Instead, they use solenoids—coil-type switches—that are triggered into action electrically or electronically.

## THOSE GREAT CONTROLS

The controls on the face of an open-reel machine are dear to the heart of its owner, enabling the person to do all sorts of things with the unit. The actual number of controls depends much on the purchase price of the machine. Less expensive models, of course, are bound to have fewer controls than their costlier relatives. Top-of-the-line units sport an array of buttons, keys, switches, and dials that can be downright dazzling.

However, regardless of price, you can expect to find certain controls on an open-reel recorder. They're pictured in the next illustration. Let's try them out and see what they do.

*Power switch* (Figure A). No explanation is needed for this one. The switch opens the way for electrical power to enter the unit.

*Function keys* (Figure B). A push of one of these starts the unit working in a certain way. Each key is marked

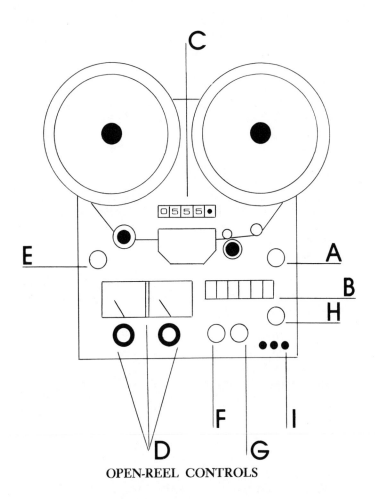

**OPEN-REEL CONTROLS**

with a word or a symbol (or both) that denotes its purpose. The symbols are employed by some manufacturers as an international code and are meant to enable people who speak different languages to use the same machine without difficulty. Practically every unit has at least five or six function keys. They are:

Record. Obviously this will be your choice when you want to make a recording. Its symbol is usually a red dot. Sometimes the entire key is colored red, to set it apart from its companions so that it won't be pushed by mistake. Activating the erase head as well as the tape head, it's the one key that you really want to avoid when intending to play back a recording.

*Playback.* This key is often simply marked *play.* Its coded symbol is a single arrowhead—>.

*Fast rewind.* Here's the key for returning the tape to the supply reel at high speed. Often marked *Review,* its symbol is a double arrowhead—<<—pointing in the direction of the supply reel.

*Fast forward.* And now the one for moving the tape onto the take-up reel at high speed. Its coded symbol is >>. Often, it's marked with the word *Cue.*

*Stop.* As the name indicates, this key halts the action of the unit. The electric power remains on, but the amplification system shuts down, and the pinch roller swings away from the capstan. The key is often marked with a blue dot.

*Pause.* Thanks to this handy little control (its symbol usually consists of two vertical lines—∥) you can halt a recording or a playback without shutting down the amplification system. It's especially helpful when you want to stop the recording process and then start again without picking up the clicks (and putting them on the tape) that always accompany the off-on actions of the amplification system. And suppose that you have to stop the recording of some radio music while a commercial is being aired; the pause key enables you to start things very quickly when the music resumes. When pushed in, the key locks itself in place and remains locked until another push releases it.

*Tape counter* (Figure C). You'll often hear this device called an index counter or a digital counter. It looks much like the odometer on a car dashboard, with the numbers rolling neatly past its windows as the reels turn. In fact, it is to the recording unit what the odometer is to the automobile, its job being to show you how much tape has traveled past the head. You'll find it especially useful for *referencing*—the recordist's word for noting the points on the tape that you want to find at a later date. Perhaps, of all the musical selections that you've recorded, your favorite is buried somewhere near the middle of the tape. You can get to it quickly with *fast forward*, if you know that it begins at the point where the counter reads, say, 232.

Shown in the illustration is a 4-digit counter. Some units have 3-digit counters; on reaching the 1,000-mark, their digits pass through zero and begin to count all over again. All counters are equipped with a control—usually a push button—that returns the digits to zero at any time you wish.

Some counters are connected to the take-up reel and others to the capstan. When connected to the reel, the counter does not measure the passage of the tape in inches or feet. Rather, it posts the number of times that the tape revolves around the core of the reel. At the start of a recording, when there is little tape on the reel, you'll see the digits flash past their windows very quickly. The pace slows as the tape gathers on the reel and increases in diameter.

Counters that are connected to the capstan measure the moving tape in inches or feet. They're usually to be found in the more expensive recorders and decks.

A caution if the counter in your unit is the take-up reel type: Because the connection is made by means of a belt, there is bound to be some slippage while the machine is running, and so you shouldn't expect the counter to be right on the nose so far as accuracy is concerned. But there's no need to worry. There will still be enough accuracy for easy referencing. When looking for that favorite musical selection, you'll usually be only a second or two away from its start on arriving at the proper count.

*Level meters and recording level controls* (Figure D). The strength of the signals reaching the tape head must always be carefully regulated. This is because no head is able to accept any and every signal that comes its way. It welcomes only those that are within a certain range of strength. If the signals are too strong or too weak, there's going to be trouble.

Should you allow too strong a signal to enter the head, it will cause *distortion* on the tape, meaning that the music or voice will sound mushy or slurred. The presence of a too-strong signal is known as *over-recording* or *over-modulation*. Almost as bad is *under-recording*, the problem of letting too weak a signal in. On playback, you'll probably hear what was recorded, yes, but it will be uncomfortably faint. And it may well be clouded over with the rumbling noises made by the unit as it's running and with the hiss of the tape as it travels past the head. All units pick up these unwanted sounds, but they're covered over and can't be heard when the recording is made at a proper level.

The recording level controls enable you to adjust the incoming signals so that they're always kept at a proper strength. The level meters show you the actual strength of the signals.

The level meters are also known as volume unit meters and are usually referred to simply as VU meters. There are at least two of them in any machine capable of making a stereo recording. Each meter is assigned to a track of its own in the tape.

A needle in each meter registers the various signal strengths by rising and falling along a scale of numbers. The scale is almost always calibrated in decibels. At one point along the way, the needle passes a warning line (often colored red) and enters a colored or shaded area. Once there, it's in the dangerous territory of over-modulation.

As for the recording level controls, each is assigned a specific job. One regulates the sounds coming from microphones. The other—called the line level control—takes care of material being piped in from a radio, a phonograph, or another recording unit.

Each control is able to cover both tracks in the tape. Making this possible is an adjustable ring, much like the focusing ring on a camera lens, that circles each control knob. By turning the ring, you can control the sound reaching one track. A turn of the knob itself regulates the sound going to the other track.

The controls shown in the illustration are of the rotary type, meaning simply that they are knobs that can be rotated. Some machines feature lever-type controls that slide back and forth.

*Tape speed switch* (Figure E). So that varying amounts of material can be put on them, tapes can be made to run through the machine at different speeds, and this is the control that enables you to select one of those speeds. Most open-reel units for amateur or semiprofessional use are built to run at two speeds—7½ and 3¾ inches per second (ips). Some units also permit a slow 1⅞ ips, and some others even have a snail-paced $^{15}/_{16}$ ips.

The ability to run a tape at different speeds is one of

the factors that makes the open-reel recorder an expensive unit. We'll see the advantages offered by the various speeds later in this chapter, when we come to the section on open-reel tapes.

*Equalization switch* (Figure F). As you already know, tape makes a hissing noise as it passes the head. This is a high frequency noise that records itself onto the tape. In a nutshell, the equalization (Eq) switch "hides" that sound. It does so, first, by strengthening the high frequency portions of the material being recorded. Then, on playback, it weakens those high frequency portions and, at the same time, weakens the unwanted hiss to the point where it is no longer heard.

*Bias switch* (Figure G). The magnetic flow going to the tape is usually weaker than the electric current that produces it. This magnetic weakness varies at different frequencies and, in the presence of a too-strong current, can cause distortion on the tape. Triggered by the bias switch is an ultra high frequency signal that takes care of the problem by strengthening the magnetic pattern being implanted on the tape. Called a bias signal, it must be of a frequency that is at least four times higher than the frequency of the highest sounds being recorded. If the frequency of the highest sounds is 20,000 Hz, then the bias signal must be 80,000 Hz. On playback, the bias signal cannot be detected because it is beyond the range of the human ear.

*Output control* (Figure H). This control governs the amount of signal that is sent out from the unit. It works in either of two ways, depending on whether your unit is a recorder or a deck. In a recorder, it is used to regulate the volume and the tone of your recording during playback.

You adjust it for loudness and for bringing out the bass and treble tones as you wish. There are usually two or more volume-tone controls on a recorder.

In a deck, the control is used to adjust the signal that is passed to the outside amplification system. A properly adjusted signal helps the amplification system send clear, rich sounds to the loudspeakers.

*Microphone and headphone jacks* (Figure I). Just as their names indicate, these circular openings permit you to connect microphones and a set of headphones to the unit. Most, if not all, units have jacks for two microphones and one for the headphones.

*Other controls.* The controls that we've talked about so far are the ones you'll find on practically all of today's open-reel recorders and decks. Some units, especially the more expensive ones, also feature a number of additional controls. For instance, as you explore the stereo shops in your neighborhood, you'll likely come across units with the following controls:

• An automatic shutoff for closing down the unit at the end of a tape. The shutoff works for both recording and playback.

• A remote control enabling you to operate the unit without ever leaving your easy chair.

• An automatic timer for turning everything on so that a radio program can be recorded while you're away. The control can also be set to trigger a playback at a desired time—a nice plus for anyone who likes to wake up in the morning to the sound of music.

Most units today also feature controls for what is known as the Dolby noise reduction (NR) system. Named

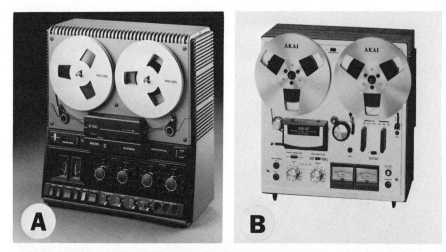

## A VARIETY OF OPEN-REEL UNITS

Today's manufacturers are producing open-reel units of varied designs. The units in Pictures A and B carry their own loudspeakers. Picture C shows a compactly built deck. (*Photo A courtesy Philips Audio Video Systems Corporation; Photo B courtesy Akai America, Ltd.; Photo C courtesy U.S. Pioneer Electronics Corporation*) for its inventor, Dr. Ray Dolby of England, the system eliminates such unwanted noise as tape hiss and recorder rumble. The system works by boosting the signal level of the low passages in a piece of music. Then, when unwanted noise makes its way onto the tape, it is covered over by the boosted passages; the high passages are left untouched because they have the power to cover the noise without any outside help. On playback, the boosted passages are pushed back down to their original levels, with the unwanted noise simultaneously being pushed down beyond the point where it can be detected.

Finally, the controls on many of today's open-reel decks and recorders are fitted with special electronic circuitry. It enables you to switch back and forth among the unit's various functions without first making everything come to a stop. For instance, suppose that you're advancing the tape on *fast forward* to the spot where your favorite musical selection is located. On arrival, you need simply press the *play* key. By itself, the machine will change speed, and in a second or so, you'll hear the music.

## OPEN-REEL TAPE

Years ago, when the first open-reel machines were manufactured, they could run at only one speed. They sent the tape past the head at a steady rate of 15 ips. Since then, as you know, the units have been improved so that they can also operate at reduced speeds of 7½, 3¾, 1⅞, and $^{15}/_{16}$ inches per second. These lower speeds were developed, of course, to enable the recordist to put as much recorded material as possible on a tape.

The high speeds—15 and 7½ ips—are the best ones for recording music. The 15-ips speed gives the finest results of all and so is used by professional recording studios and by amateurs who want the ultimate in sound quality. The

## PROFESSIONAL OPEN-REEL UNIT

The marked difference between home and professional open-reel units can be seen at a glance in the ATR-124 multitrack recorder from the Ampex Corporation. The unit is capable of recording 8, 16, and 24 tracks. (*Photo courtesy Ampex Corporation*)

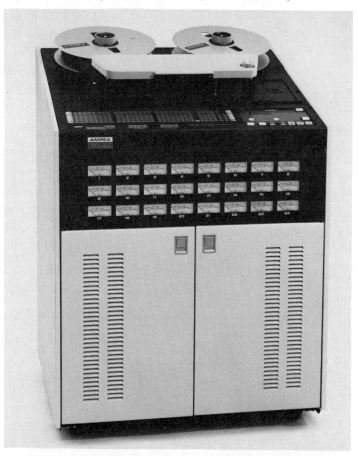

3¾ and 1⅞ speeds should be reserved for recording the voice. As for the ¹⁵⁄₁₆ speed, it's really intended for nothing but the voice.

Why is it best to record music at the higher speeds? Because, with a greater length of tape passing the head each second, the magnetic flow has a larger area on which to imprint the magnetic patterns. The patterns end up being very complete and catching all the tones in the sound. The result is a rich and full recording.

That same result can't be obtained at the slower speeds. With the magnetism having less space to work in per second, some of the tones won't be caught and implanted on the tape. On playback, you're likely to hear some pretty "tinny" sounding music.

The speaking voice is also best recorded at the higher speeds, and for the same reasons. But, because the voice doesn't give off such a complex array of sounds as does music, it can be quite successfully recorded at 3¾ ips. The speeds below 3¾ don't really do a good voice job and are available mainly for convenience. They can be used without using up a lot of tape at those times when voice quality isn't important. For instance, when recording your friends' joking at a party, you won't be too worried about how great they sound; you'll be more interested in what they're saying and in getting as much material on the tape as you can. The same is true if you're dictating a story or some class notes into your machine.

Open-reel tape for amateur use is usually one-fourth of an inch wide. To give you a variety of recording times, it can be purchased in several lengths. You'll find it for sale on reels that measure 5, 7, or 10½ inches in diameter.

The 5-inch reel is used by many home recordists, but the 7-inch one is a far more popular choice. It can hold tape lengths of 1,200, 1,800, and 2,400 feet. Traveling in one direction at 7½ ips, these tapes will give you 32, 48, or 64 minutes of recording time respectively. The times, of course, are doubled if the tapes are run in both directions.

The thickness of the tape determines the length that can be put on the reel. The 1,200- and 1,800-foot tapes are one millimeter thick, whereas the 2,400-foot stretch measures just half a millimeter. The 2,400-foot tape may seem appealing because it offers the greatest amount of listening time, but most recordists don't like it. The problem is that it's thin and can be too easily stretched or torn. And it's vulnerable to the headache known as *print-through*.

Print-through is always a threat when you store a tape for a time after making a recording. The magnetic patterns somehow work their way through the acetate base and imprint themselves on the adjacent layers of tape. The result: An echolike sound intrudes itself on the music when the tape is next played. The thinner the tape, the greater is the danger of print-through.

If you plan to record at 15 ips, you'll be wise to buy your tape on a 10½-inch reel. Though only about a third larger in diameter than the 7-inch reel, it holds twice as much tape. This is because of the wide circumference that the tape reaches as it is being wound onto the reel in those final three inches.

As was said at the beginning of the chapter, the open-reel recorder is the most expensive unit that you can buy, thanks to its many features, its various tape speeds, and the fine engineering that has gone into it. Should you become a serious recordist, you'll probably want to use it because of the top-quality sound it gives. Also, with the tape open to your touch, you'll be able to do plenty of editing and splicing work, a job that is difficult to do with cassette tape and sometimes impossible with 8-track tape.

But it may take you quite a while to save the purchase price of an open-reel machine. Remember, those prices run from a low of about $500 to more than $2,000. In the meantime, you needn't lose out on all the fun of recording. There's much pleasure—and much excellent sound—to be had for far less money if you turn to either the cassette or the 8-track.

Let's get to know them.

CHAPTER

# ABOUT CASSETTES
# AND 8-TRACKS

THE CASSETTE and 8-track are very popular today because of their size and price.

Compactly built, both take up less space than the open-reel unit and its somewhat bulky tapes, a fact that appeals to recording fans who live in smaller homes, apartments, and school dorms. And both are nowhere near the open-reel in price. Good cassette and 8-track decks can be purchased for around $200 and, if you keep your eyes open for store sales, you can probably find something for well below that figure. A portable cassette recorder (Picture A in the illustration below) usually costs between $50 and $100.

Careful listeners say that the sound produced by the cassette and the 8-track is not up to open-reel quality. There's no doubt that this is true, but there's also no doubt that the two units, and especially the cassette, produce good sound. Cassette sound can be so close to open-reel quality that the average person will have difficulty in telling the difference.

## ABOUT THE CASSETTE

Pictured in the next illustration are three different cassette units. Pictures A and B show two portable recorders, with one of them containing an AM/FM radio. Both units operate on small batteries so that they can be used anywhere, and both come with a cord attachment for tying them into wall outlets, should you wish to use electric power. Both have built-in microphones; to record your speaking voice, all you need do is talk in the direction of the recorder. And, finally, both have built-in speakers for playback.

Picture C shows a cassette deck. If you look closely at the deck, you'll see that it's equipped with the same controls as the open-reel unit. Included are the power switch, function keys, bias and equalization controls, tape counter, VU meters and recording level controls, and jacks for microphones and a set of headphones.

The portable units are equipped with many of these same controls. Most portables—to keep their size and price down—are not outfitted with VU meters. Rather, they come with a small red light that flashes a warning when you begin to over-record. In place of the recording level control, they have either a lever for adjusting the level or a built-in control that automatically regulates it.

It so happens that each of the two portables in the

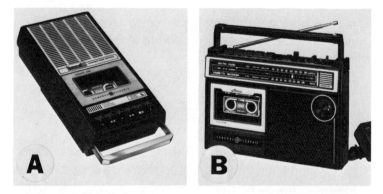

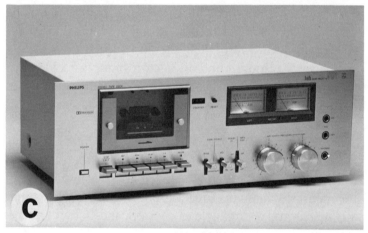

## THREE CASSETTE UNITS

Two portable recorders (Pictures A and B) and a deck (Picture C) are seen here as examples of the cassette units being produced today. The recorder in Picture B contains an AM/FM radio. (*Photos A and B courtesy General Electric Company; Photo C courtesy Philips Audio Video Systems Corporation*)

illustration has a digital tape counter. This is a feature that less expensive units often go without. When using a counterless model, you can keep track of things by checking a series of reference lines that are inscribed on the plastic tape case or on the wall of the tape compartment in the unit. Even machines that come with counters have these markings in their compartments. By looking closely, you can see them in each of the pictured units.

The tape compartment is a feature that all cassette units have in common. It is located behind a metal-framed glass door and contains two spindles onto which the tape case is fitted for recording or playback. A function key marked *eject* is used to open the door. On most units, the same key works both the *eject* and *stop* functions. A certain amount of finger pressure brings the tape to a halt. A little added pressure animates spring hinges on the door and pops it open.

An illustration on the next page shows inside the unit. As shown in Figure A, a cassette is equipped with a capstan and a pinch roller. Just as in the open-reel, the capstan is made of a fine quality steel, and the pinch roller of a soft rubber. Also to be seen are at least two heads—one that combines recording and playback and another that handles erase. Some expensive units, of course, have three heads, with separate heads taking care of recording and playback.

Because the tape is contained inside a plastic case, one question is always asked by beginning recordists: How do the heads, the capstan, and the pinch roller all manage to make contact with it?

The answer begins with the design of the plastic case itself. If you look down on it from above (Figure B), you'll

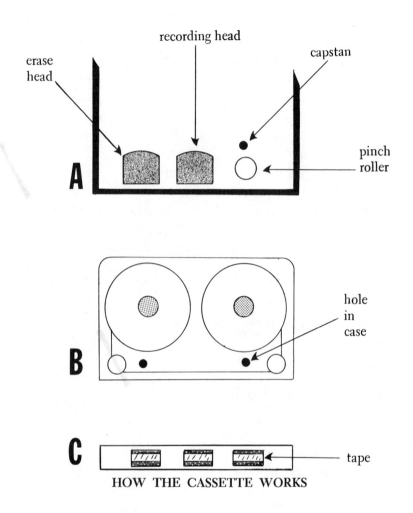

erase head

recording head

capstan

pinch roller

**A**

hole in case

**B**

**C** tape

**HOW THE CASSETTE WORKS**

see the tape inside coming from the supply reel, curving past the guide roller, and then traveling along the base until it reaches another guide roller at the opposite end. There, it turns and runs to the take-up reel. Near the base you'll see several holes in the plastic (just two are shown in the illustration). They extend all the way through the case.

Now let's turn the case on end (Figure C) and look along the length of its narrow base. You'll see that there are three rectangular openings in the plastic. If you put your little finger in any one of them, you'll be able to catch the tape with your fingernail and pull out a bit of it in a loop.

The openings are carefully spaced. When you fit the case into the unit, the center opening will line itself up with the recording-playback head. The opening on your left will drop in opposite the erase head, and the opening on your right will end up opposite the pinch roller. As for the capstan, it will extend up through the hole behind the right opening.

When you first fit the case into the unit, the recording head and the pinch roller will be a little distance from the tape. But now let's switch the unit on and begin the recording. Both the head and the pinch roller are pulled forward by a spring device and come to rest against the tape, the head touching it lightly and the roller pressing it against the capstan. At the same time, an electric motor springs to life inside the unit and begins to turn the capstan and the take-up reel. The tape brushes past the head at that steady pace needed to give you a good recording.

Many of today's cassette units use just one electric motor. It works in the same way as the open-reel motor, moving the tape along at a steady recording and playback speed and then racing it in one direction or the other on *fast rewind* and *fast forward*. More expensive machines use two or three motors. In the two-motor unit, one motor usu-

**MINIATURE CASSETTE RECORDER**
Modern technology has made possible the production of miniature recorders. Here's one that will fit right into your hand. (*Photo courtesy Radio Shack*)

ally operates the capstan while the other attends to the reels. In the three-motor unit, of course, the capstan and the reels have their own motors.

## ABOUT CASSETTE TAPES

So that it will take up as little space as possible, cassette tape differs from open-reel tape in several ways. First, all cassette reels are of the same size—a diminutive two inches in diameter. They all hold a tape that is only one-eighth of an inch wide, just half the width of open-reel tape. Yet, even in this narrow space, it's divided into four tracks so that a stereo recording can be made in both directions.

Next, cassette tape can be made to run at only one speed—a slow 1⅞ ips. The early developers of cassettes settled on this speed for the obvious reason that it allows a great deal of recorded material to be put on those tiny reels.

The slow speed may cause you to lift an eyebrow because you know that 1⅞ ips on an open-reel won't give you a good musical sound. The fact is that when the cassette was first marketed, its sound quality was nowhere near that produced by the open-reel. But the early manufacturers knew that their product would never achieve widespread popularity until the problem was solved. And so, over the years, some intense work was done to improve both the unit and its tape. One of the great tape improvements came with the development of a more sensitive magnetic coating. Today, as was said at the beginning of the chapter, cassette tape has come so close to its open-reel cousin in sound quality that the average listener really can't detect the difference between the two.

Unlike its cousin, cassette tape is not packaged in foot lengths. Rather, it comes in playing-time lengths. These lengths are usually divisible by 15 minutes. And so, when you're ready to buy your first tapes, you'll find that some cases are labeled C-30 and others C-45 (the C, of course, stands for cassette), meaning that the first will run for 30 minutes when played in both directions, and the second for 45 minutes. Others are labeled C-60, C-90, and C-120.

Cassette tapes are not as thick as open-reel tapes. A 1,200-foot open-reel tape, you'll remember, is one millimeter thick, but the cassette measures only one-half millimeter thick in its C-30 to C-60 lengths. The thickness goes down to 0.3-millimeter on C-90 tapes, and to a hairbreadth 0.25-millimeter on C-120s.

Unless you feel it's absolutely necessary to use them, you'll be wise to avoid the C-90 and C-120 lengths. Thin as they are, they can cause all sorts of trouble. The danger is great that they'll jam while traveling from reel to reel. Even greater is the danger of breakage. And, when stored with recorded material on them, they're especialy susceptible to the headache of print-through.

The final difference between cassette and open-reel tape is the most obvious one of all. You can touch and handle open-reel tape with no trouble whatsoever. But cassette tape, locked within its plastic case, is difficult to reach. Consequently—to repeat a point made in Chapter Two—open-reel tape is much easier to deal with when you want to do editing or splicing. In fact, many recordists advise that you go for the open-reel unit and forget about the cassette if editing and splicing are to be major activities in your hobby. These enthusiasts feel that it's best to leave the cassette tape alone unless it's in need of repair.

Now let's talk about a handy feature that's found in all cassette tape cases. Suppose that you make a recording of your own or buy a commercially recorded tape containing some favorite music. Whenever you play the tape, there's always the danger that you'll accidentally press the *record* key and activate the erase head. Some great sound is going to be wiped out by the time you realize your mistake.

A simple device shown in the illustration opposite removes the danger for good. Called an erase prevention mechanism, it consists of two little plastic tabs on the narrow side of the case opposite the recording base. They're located, as seen in Figure A, right at the corners. When

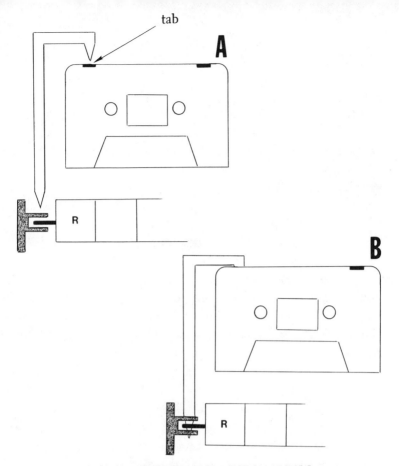

tab

**A**

**B**

**ERASE PREVENTION MECHANISM**

they're in place, you can make a recording on the tape. But they can be punched out with a penknife or a small screw-driver and, once this is done, the tapes can be used only for playback.

Figures A and B illustrate how the erase prevention mechanism works. On being placed in the machine, the tape case drops in alongside an L-shaped pin. When the tab is in place (Figure A), one prong of the pin rests against it, preventing the long stem of the L from moving forward and inserting itself in a locking bar on the *record* key. But if the tab has been removed (Figure B), the prong sinks into the case. The long stem then slides forward and inserts itself into the locking bar, making it impossible for anyone to activate the *record* key. The tape is safe from accidental erasure.

Though we've talked about just one tab in the explanation, remember that there are two in each case. One is used to protect the first side of the tape. Its companion, of course, protects the second side.

Now suppose that, after punching out the tabs, you decide to make a new recording on the tape. All that you need to do is place little strips of plastic tape over each tab. They'll be strong enough to prevent the prong from entering the case, and you'll be all set to go.

## ABOUT THE 8-TRACK

As was mentioned in Chapter One, the 8-track unit was first developed for drivers who wanted to listen to commercially recorded stereo music while making their way through traffic or along the highway. In the car, it is a small player—not a recorder-player—that is usually attached to the underside of the dashboard. If you buy a unit for your car, you'll need to install two loudspeakers for stereo sound; some cars, however, come with the player and speakers built in.

The tape itself is housed in a plastic cartridge that measures four inches wide by five inches deep. There are openings in one end of the cartridge that permit the tape to come into contact with the tape head. The tape has,

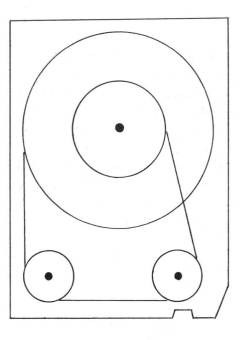

**8-TRACK TAPE
IN CARTRIDGE**

you'll recall, the amazing capacity to play itself over and over again without ever being touched.

The tape is able to perform this feat because it is wrapped in a continuous loop around a single reel in the cartridge. When moving, the tape rises out of the center of the cartridge, curves its way around two guide rollers, passing the head en route, and returns to the outside of the reel. From there, it works its way back to the center so that it can start its trip all over again.

The capacity to play endlessly has made the 8-track tape a delight for drivers because it is so safe and convenient to operate. Because there's no need ever to turn the tape over, drivers don't have to let go of the steering wheel every few minutes or take their eyes off the road. They merely switch the unit on, slip a favorite cartridge into place, and let the tape do the rest.

Lined along its length with eight tracks, the tape easily produces stereo sound. The head, equipped with the customary two recording gaps for stereo pick-up, takes the sounds from two tracks at a time. But, to cover all eight tracks, it does something that makes it different from all other heads. It moves up and down at times during the playing process, riding from one set of tracks to the other. Using the illustration below, let's follow its travels.

## 8-TRACK HEAD AND TAPE

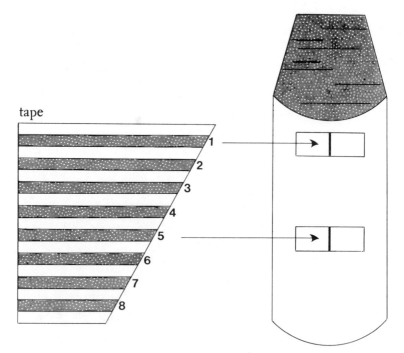

tape

At the start of the tape, the recording gaps are aligned with tracks 1 and 5. The tape plays its way through to the end of the tracks. At that point, a metal strip on the tape makes contact with a switching device in the unit. This electrical contact triggers the head into moving downward until the gaps align themselves with tracks 2 and 6, the next ones to be played. When they finish their run, the head is again triggered into moving downward, this time to tracks 3 and 7, and then, a little later, to tracks 4 and 8.

Once tracks 4 and 8 have played themselves out, the metal strip at their end makes contact with the switching unit and the head rises back to tracks 1 and 5. The cycle of playing starts all over again. The cycle continues to repeat itself—with the driver listening to the same musical selections again and again—until the unit is switched off.

Each pair of tracks usually takes about fifteen minutes to run past the head. All four sets, then, give about an hour's worth of listening time before the musical selections begin to be heard again.

There is a pause of several seconds each time the head moves from one set of tracks to another. Because of this pause, most commercially recorded 8-tracks contain popular or light classical musical selections, each rather brief so that several can be fitted completely into each fifteen-minute segment. Not many symphonies are recorded on 8-tracks. Symphonic movements often run too long for the segments, and manufacturers don't like to intrude in the middle of a movement, knowing that the pause may spoil the entire symphony for the lover of serious music.

Because 8-track units proved to be popular in cars, manufacturers soon developed outfits for home use so that people could get double use out of their tapes. Today, most of the 8-track units used in the home are players only. Only a few models also make recordings.

The 8-track units have never been widely used as recording machines because the single-reel tape is difficult to rewind or advance when you want to make a quick check of recorded material. In fact, the tape is practically impossible to rewind; as a number of stereo buffs have pointed out, it's about as easy to rewind an 8-track tape as it is to put toothpaste back inside a tube. So far as advancing the tape is concerned, some units do have a *fast forward* key. But the tape advances at only a fraction of the speed attained by open-reels and cassettes.

Though without a *fast rewind*, the 8-track recording deck contains most of the controls that are found on a cassette of similar quality. In all, the unit looks much like a cassette at first glance. The most obvious difference is the compartment in which the cartridge is housed. Rectangular in shape, it is just wide enough from top to bottom to accept the cartridge, face up. The compartment is covered with a little door, made either of plastic or metal, that opens inward and upward when the cartridge is pushed against it. Once in place, the cartridge protrudes slightly out through the doorway.

## ABOUT 8-TRACK TAPE—SOME PROBLEMS

An 8-track tape is one-fourth inch wide and plays (and records) at 3¾ ips. Though as wide as open-reel tape and traveling faster than cassette tape, it produces a sound that is somewhat below theirs in quality. This is due to several factors.

First, there's the fact that the tape is divided into eight tracks. The tracks are so narrow that the moving head often has trouble in precisely aligning its gaps with the tracks. A certain amount of noise rises behind the music when the alignment is even slightly off. Perhaps the problem could be solved by widening the tape, but then the unit would sacrifice the compactness that enables it to fit so conveniently into the dashboard of an automobile.

Another problem is caused by the single-reel design. As the tape is being pulled out of the reel hub, it is subjected to a great deal of friction. This friction can cause changes in speed that produce those bugaboos called wow and flutter.

Still another problem rises out of the manner in which commercially recorded 8-track tapes are duplicated for mass sale. They're duplicated at very high speeds, a process that results in more background noise than can be detected on cassette and open-reel tapes. In a car, this noise is usually hidden by engine, wind, and road sounds. But it can be a problem for the careful listener when the tape is played in the quiet of the home.

Despite these problems, it must be said that a great many people are quite satisfied with the quality of sound produced by their home units. If you own an 8-track and are pleased with it, then don't worry about anything that has been said here. What counts is that you are enjoying the set.

Now that we've taken a close look at the various recorders and decks, it's time to take the next step. It's time to make your first recording.

# YOUR FIRST RECORDING

THERE ARE two types of recordings that you can make: *donor* and *live*. In donor recording, the material comes from (is donated by) a phonograph, a radio, a television set, or another recording unit. In live recording, of course, the material comes from people who are in the room with you. If you wish, you can blend the two types together. With experience, for instance, there's nothing to stop you from recording a friend's voice with phonograph music in the background, just as is done in radio and TV dramas.

## DONOR RECORDING

If your recording unit is part of a stereo set, you'll have no trouble making your first recordings. The unit is already connected to some donors—usually a phonograph and an

AM/FM receiver—and is ready to work at the flick of a switch. But what if it isn't part of a stereo set? Before you can make a recording, you'll have to choose a donor and tie the unit into it.

Right away, let's talk about how to make that tie-in.

The simplest way to do the job is to place a microphone in front of the donor's loudspeaker and make the recording through it. Unfortunately, this is also the least satisfactory way of doing things. There's bound to be some distortion in the sound that you pick up, especially if your donor set is on the inexpensive side and doesn't produce a sharp and clear sound to begin with, or if it's so old that it's no longer working up to par. You're almost certain to get a recording that sounds pretty muddy.

Worse yet, the mike is going to pick up many unwanted noises from around the room. Suppose that your older brother, who always sounds as if he has size 42 feet, comes clomping in from the kitchen. Or your dog arrives and greets you with a few happy barks. Or the telephone rings. And there you are with a recording that's made up of nice music, *clomp-clomp*, more nice music, *yap-yap*, still more nice music, and then *jangle-jangle*. It's enough to discourage anyone.

If you can get through the session without encountering these problems, you may end up with a fair recording, especially if it's a spoken-voice recording that you want for practical purposes rather than for quality of sound. For instance, the mike-in-front-of-the-donor system will serve well for recording a presidential speech that's to be used in a report for your social studies class. But, if it's listenable music that you're after, you're going to be disappointed.

Incidentally, should you run into trouble with background noise, you might try wrapping a handkerchief or small towel around the body of the microphone. Leave the mouthpiece uncovered, of course. The thickness of the padding should muffle some of the sounds and eliminate others altogether.

There are two other ways to make the tie-in. Both will give you a better recording. They both require that you remove the back panel from the donor.

Then you may tie the recording unit to the donor's loudspeaker terminals. Or you can tie into the donor's volume control terminals. Each method reduces the distortion problem, with the volume control tie-in cutting it out completely.

These tie-ins, though they're not particularly difficult to make, should never—but *never*—be attempted by anyone who isn't expert in handling electrical equipment. There's always the danger of electrical shock, particularly in AC-DC donors; they frequently carry voltage in the chassis and can really knock you back on your heels. And there's the danger of making an incorrect connection that will damage the donor or the recording unit, or both.

And so, if you must make either of the tie-ins, be sure to seek help from someone with electrical experience. Perhaps a knowledgeable friend or family member can guide you. Or the teacher in your electrical or electronics shop at school. Or the technician at your neighborhood stereo shop. Above all, don't just open the donor up and begin experimenting. That's really inviting all sorts of trouble.

All right. Let's assume that you've made the tie-in or that your unit is part of a stereo set. One way or the other, you're ready to go. There are several steps that you should now follow in making a good recording. We'll give them a try now as we make recordings from a phonograph, then from a radio broadcast, and finally from another recording unit.

## RECORDING FROM A PHONOGRAPH

Your first job is to check the condition of the disc to be recorded and to make sure that it's clean. Dust specks, oily streaks from your fingers, and stains of any sort can all cause clicks and distortions in your recording. No matter how clean the record looks, it's a good idea to wipe it off with an antistatic cloth. If the disc seems at all dirty, you'd best treat it with a cleaner. Antistatic cloth and cleaning kits can be purchased for a few dollars at any stereo shop. Be sure to follow the directions for use that come with both.

The next job is to adjust the VU (volume unit) meters to the point where they will ensure the best recording possible. In particular, you want to keep them from bouncing too much into the danger area that will cause distortions in the sound.

**VU METER**

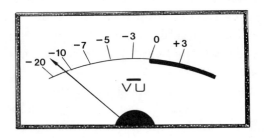

As you know, you enter the danger area at the point where the meters are marked 0. But there's something more that you need to know. Some manufacturers give you a margin of safety by calibrating the meters so that you don't actually run into distortion trouble right away. But other manufacturers hardly give you any leeway at all. There's only one way to find where distortion actually begins in your unit, and that's to put it to a test.

Before turning the unit on, play your record through and pick a passage that contains loud, medium, and low sounds. Now switch the unit on, but don't set the reels to turning just yet. Rather, play the passage through and watch the needles in the meters. Adjust them so that they never pass the minus 5 mark on the scales. If you have a Dolby noise reduction system in your deck, be sure that it's turned off so that a full range of sound will be present, along with all background noise such as tape hiss.

Now, with the reels turning, make a recording of the passage, not letting the needles sail past that minus 5 mark. at the end of the passage, let the tape continue to run. Reset the phonograph tone arm and record the passage again, this time adjusting the meters so that the needles *peak out* at 0 on the scales.

Without stopping the tape, record the passage once more, but now adjust the meters so that the needles peak out above 3 or a couple of divisions beyond the 0. Then it's time for a fourth recording, with the meters adjusted so that the needles bounce clear over to the very ends of the danger zone.

And that's it. The recording part of the test is finished. You're ready for playback, once with the Dolby system turned off and once with it turned on. Send the tape through and listen to each recording with the volume control set so that the softest music in the passage can be heard comfortably in a quiet room. Listen carefully. Can you detect tape hiss or recorder rumble at the minus 5 setting? Or does the music seem muffled at times? Is there distortion at the 0, plus 3, or maximum setting?

A few minutes of listening, or several replays of the tape, should give you a pretty good idea of a safe meter range for the actual recording—a range that picks up the soft sounds nicely and avoids distorting the loud ones. Suppose that you begin to distort at 0 or plus 3. Then adjust the level so that the needles will peak at or just below these points. Once you've made the adjustment, leave the meters alone. Do not adjust them during the actual recording.

The test is a time-consuming one to make, but it's a necessary one, so be patient and see it through. The chances are that you won't have to repeat it every time you make a recording in the future. Before long, your growing experience as a recordist will give you a feel for proper meter adjustment. Very likely, you'll end up listening to the disc and knowing instinctively where the meter should be set.

Now it's time to make the actual recording. But wait a moment. Here you can run into an annoying little problem, especially if you're working with a phonograph that has an automatic changer. If you start the disc automatically, there's going to be a thumping sound as the tone arm rises out of its cradle, and then a definite click when the stylus (needle) settles on the disc and heads for the grooves. Both sounds are going to make their way onto the recording.

They can be easily avoided. First, press the *record* key to get the tape moving and then go to the *pause* key. Now, with the tape stopped dead in its tracks, start the phonograph. Let the tone arm swing over and drop to the disc. Once the needle has settled and is moving toward the first groove, release the *pause* key. The music will start free of any unwanted noises.

There will also be a clicking problem at the end of the disc if you're working with an automatic changer. It comes when the tone arm rises from the record for the return to the cradle. Again, the *pause* key is the answer. As soon as the music ends, press the key. Then, if you have a second disc on the changer, let the key remain depressed until the new record drops into place and the stylus settles into place.

Even if your phonograph doesn't operate automatically, you should use this same technique at the start and close of a disc. There's bound to be a click when your hand lowers the tone arm to the disc, and then a swishing sound at the end when the stylus moves into those widely spaced grooves that carry it to the turntable's spindle. Incidentally, if you have a damper on your turntable, be sure to use it instead of your hand when putting the stylus in place or taking it away. The damper is a control that raises and lowers the stylus from the disc. It does so without a click or, at the least, with a very tiny, hardly audible sound.

Now what can you do if your recording unit doesn't have a *pause* control? Again, set the tape to moving, but with your level controls turned all the way down so that no sound can come through. Let the stylus land and then, as it heads for the opening groove, bring the controls up to the proper recording level. Bring them up quickly and smoothly. If you bring them up too slowly, the stylus will beat you to the music and the first notes won't be fully audible. They'll fade in as if the musicians are marching in from a distance. At the end of the disc, of course, run the controls back down as far as they'll go.

Also, if you're working without a *pause* key, it's a good idea to let the tape go on running while the next disc drops into place. You'll use up some extra tape this way, but you'll avoid the clicking that is heard whenever a unit is switched off and on. Later, you can edit the dead sections out.

A recording tip: When using a cassette, you'll be wise to memorize the exact numbers on your counter at which the various tape lengths come to an end. Because cassette tapes are measured in playing times rather than foot-lengths, you can't glance at the counter and then quickly subtract its current number from the known foot-length to see how much tape is left. You have to know beforehand the numbers on which the C-30, C-60, and other lengths run out. This knowledge will prove especially helpful in the recording of serious music, such as an opera or a symphony. For best listening, you'll want to turn the tape over between scenes in an opera (or at least between major arias) and between movements in a symphony. Only by knowing the ending numbers will you be able to determine whether to go on to the next selection or shut down for the turn-over.

## RECORDING RADIO MATERIAL

When you record music from a radio broadcast, you don't usually have the chance to set your VU meters according to the selections to be recorded. But you needn't work totally blind either. You can help yourself by tuning in the broadcast and letting the needles bounce back and forth while you listen to music that is similar to the type you plan to record. In this way, you'll at least come up with a beginning idea of where the meters should be set. Also, remember that your growing experience will soon be giving you a feel for correctly setting the meters.

But the setting will be approximate, and so you'll have to "ride the gain" during the recording session, adjusting the level controls as you go along. Here, again, you needn't work totally blind. If you're recording popular music, the odds are that you'll have heard the selections before. You'll be able to anticipate the loud passages and lower the gain just as they come pouring through.

It's especially challenging to record music that's new to you. Then you must be doubly on guard for all the major sound changes and make your level adjustments quite swiftly. Actually, you'll be responding to the loud passages and making the adjustments a split second after they hit. Up against this sort of problem, you can't expect to stop distortion at all times. What you must aim to do is keep it at the barest minimum possible.

Whether or not the music is familiar to you, concentrate on riding the gain smoothly. Raise and lower it at a deliberate speed, never whipping the controls back and forth. Try to keep each adjustment as slight as possible, just enough to do the job for you. By operating the controls gently and expertly, you'll keep the music from fading in and out.

You'll find it much easier to regulate the levels for popular music than for classical music. This is because popular music is commercially recorded at what is called a constant volume level so that it can be used without a great deal of adjustment by various radio stations and in the countless jukeboxes found all across the country. Classical music, on the other hand, is commercially recorded in a manner that allows it full range of sounds to come through, requiring that you be especially vigilant when making tapes of it.

You'll also get some outside help when making a recording from a radio broadcast. First, the radio station itself —to avoid distortion—regulates the sounds coming through and so makes your job a little easier. Second, if your recording unit is one of the more expensive models, it may contain an automatic overload-prevention circuit. Basically, in nontechnical terms, this circuit reduces distortion by affecting the loudness level (the circuit also works when you're recording from a phonograph or another recording unit).

Recording tip: If you want to cut out the commercials and announcements that come between the musical selections, just depress the *pause* key as soon as the last note has died away. Then release it in the second or two before the music is to start again. But be prepared for the station to trick you now and again. Expecting a commercial at the end of a selection, you may hit the key only to have the station launch into another number (the chances are it will be one of your favorites). On the other hand, you may release the key at the end of the commercial and then have to put up with some more happy words from the announcer instead of music.

What to do? It's all up to you. Should music come through when you're expecting a commercial, you can immediately release the key and do without the first few bars.

Or, if you're a recordist who finds even one missed note intolerable, you can shrug, murmur "Better luck next time," and wait patiently for another selection.

And what if another commercial comes through instead of music once you've released the key? You can either depress the key again or let the tape run, planning then to edit out the unwanted voice at a later date. Or, while the commercial is on, you can back the tape up to the end of the preceding selection and let the machine erase the words when you start to record again.

Your on-the-spot solutions to these various problems are all part of the adventure of recording.

If your unit doesn't have a *pause* key, you'll need to leave the machine running or turn it off between musical selections. If you choose to leave it running, the commercials and announcements can be edited out later. Many recordists, though risking some clicks and thumps, prefer to turn everything off because the announcers' messages eat up too much precious tape.

Another tip: When recording a radio broadcast, always try to use an FM station. You'll find FM stations free of the static that bothers AM outlets now and again.

Though we've talked here only about making a tape from a radio broadcast, all the same suggestions can be applied to recording sounds from a television station.

### RECORDING FROM A FELLOW UNIT

If you're fortunate enough to own a second recording machine, or if a friend has one, you can tape material from it. All that you need do is tie the two units together. Practically every recording unit has input jacks in its rear facing that enable a fellow unit to be connected to it.

Once the connection is made, you should work just as you did when recording from a phonograph. Play a passage from the tape on the donor unit several times while you locate the proper VU settings on your machine. Once you've found the proper settings, leave them untouched as you make the actual recording.

## TENDER LOVING CARE

Once you begin to make recordings, your tapes and your unit are going to require some special care to keep them in top working order. Special though it is, the care is easy to give. All that you need do is put the following tips to use.

### TAPE CARE

The suggestions for tape care have to do with safe storage and begin with the warning that all tapes are vulnerable to a certain kind of danger. Coated as they are with magnetic particles, they can easily pick up magnetic fields coming from such gear as your amplifier and loudspeakers. The result can be a partial erasure of the recorded material. And so, right from the start, your tapes should always be stored in a spot that's well away from the rest of your recording and stereo gear.

The tapes should also be stored in a place where the

temperature is relatively even. They'll remain in fine shape for many a year if stored at a room temperature of around 70 degrees Fahrenheit, with the humidity between 40 and 60 percent. Watch out for any storage spot that might become icy cold in the winter. Extreme cold can make the polyester dangerously brittle.

Tapes should not be wound too tightly when in storage. A tight wrap, which is mainly caused by returning the tape to its starting point on *fast rewind*, stretches the polyester and, to make matters worse, adds to the possibility of print-through. You'll be wise to store your tapes as the professionals do—*tails out*. This simply means to store them without going through the rewind. They'll then be subjected to the least amount of tension. When you're ready for a playback, then it will be time for the rewind.

As protection against dust and grime, all tapes should be returned to their containers as soon as you've finished using them. For convenient storage, you can purchase plastic containers large enough to hold a dozen or more cassette tapes; they're to be found for a few dollars each at any stereo shop. If open-reel tapes are to be stored for long periods, it's best to place them in metal containers. Containers of the type used for 8-millimeter movie film are perfect choices here.

## Unit Care

The heads, capstan, and metal guide rollers in a unit must be cleaned periodically. They need attention because they're constantly being touched by the tape. As it brushes past, minute particles break away from its oxide coating and collect on them. A muddy sound results from the build-up on the heads. As for the capstan and guides, the build-up causes them to stick to the passing tape, triggering speed variations that produce flutter and wow.

For the cleaning job, you may use either isopropyl alcohol or a solvent recommended by the manufacturer (never use a solvent that is not specifically recommended in your owner's manual). Or, if you prefer, you may try one of the cleaning kits that are on sale in the stereo shop. Whatever your choice, gently rub the heads, the capstan, and the rollers with a cotton swab that has been dipped in the cleanser. The swab should come away with a brown oxide stain. If the stain seems especially heavy, let the cleanser dry (this will take only a few seconds) and then give the parts a touch with a fresh swab.

A caution: Unless the directions that come with the cleanser say otherwise, you should clean the guide rollers only if they're made of metal. Some cleansers shouldn't ever be used on rubber components because they will tend to make the rubber stick to the tape. Take particular care to keep such cleansers away from the soft rubber pinch roller.

At the least, your unit should be cleaned after every forty hours of use. It's wise to get to the job more often if you work the unit really hard. Some recordists feel that there should be a cleaning once a week if the unit is made to run from ten to fifteen hours in that time.

In time, the recording-playback head will build up a residual magnetic field—a magnetic field of its own. If left untended, the field causes increased tape hiss and a loss of the high frequencies being recorded. Whenever you suspect its presence, it should be eliminated immediately with a demagnetizer.

The demagnetizer, which is known technically as a degausser, is an expensive hand-held tool that's available at any stereo store. Built to run on house current, it features a casing with a small wire coil inside. Extending from the casing is a flat prong—the pole—whose flat tip can rest easily against the front of the heads.

Simply plug the degausser into a wall outlet, place the pole tip against the head face, and then pass the tip over the entire face for a second or so. Be sure to stand well away from the recording unit when plugging the degausser in. And pull away to a distance of about four feet before unplugging it. If you stand too close at the start or finish, you'll end up increasing rather than eliminating the magnetization in the head. Make sure, too, that all your tapes are safely out of the way. Otherwise, you're risking damage to your treasured recordings.

Should you record again and again on the same tapes, you may find that your erase head is at last unable to remove all the material from them. Here, a bulk eraser—another hand-held accessory that works on house current—will be of great help. Just by passing it over a reel you can clear the entire tape.

In addition to your cleaning and demagnetizing chores, you'll need to check the mechanical workings in your unit periodically to see how they're faring. You're going to find that such items as pads and rubber belts have to be replaced from time to time. They slowly but surely wear out with ordinary use. Unless you really know what you're doing, don't attempt any repairs and replacements on your own.

Take the unit to an expert repairman. The trip will cost you a little money, but probably far less than if you try to fix things on your own and then, despite your good intentions, make a hash of the job.

As was mentioned right at the beginning of this chapter, you can make both donor and live recordings. We'll be getting to live recordings in Chapter Six. But, before we can really talk about them, we need to get acquainted with yet another item of equipment—the all-important microphone.

CHAPTER

# MEET THE MICROPHONE

ALL LIVE recording begins with the microphone. No matter what the microphone looks like (if you thumb through some stereo catalogs, you'll see that it comes in a variety of shapes, sizes, and designs), it has just one function: It picks up the sound waves created by the voices and musical instruments near it and converts them into electrical waves.

## HOW THE MICROPHONE WORKS

Because it has the power to change sound waves into electrical ones, the microphone is known technically as a transducer. It contains two devices that, by working together, enable it to do its job. The first is the diaphragm, which is located directly behind the microphone face. The diaphragm is a thin sheet of semiflexible material that, like the eardrum, begins to vibrate whenever sound waves strike it.

The vibrations immediately pass through a mechanical connection to the second device—the transducing element. This is an electrical unit that now vibrates in perfect rhythm with the diaphragm. In so doing, it generates a small electric current. Off to the recorder or deck goes the current, where it is strengthened and then passed to the recording head.

There are several kinds of transducing elements, and a microphone is usually named for the particular element that it contains. Topping the list of elements used in today's microphones are four: the piezo, the electrodynamic, the ribbon, and the condenser. Each type generates an electrical current in its own way. With the help of the illustration on the next page, let's see how each works.

*The piezo microphone* (Figure A). The piezo element works by air pressure and is named for the ancient Greek word meaning "to press." It is made of a ceramic substance or a crystal such as rochelle salt crystal. When the element vibrates, the arrangement of its atoms is changed by the pushing of the surrounding air. This change triggers the electrical charge that is sent to the recording head.

The piezo mike is the most inexpensive of microphones. It often comes along with a recorder or deck as accessory equipment provided by the manufacturer. The changes wrought in some crystals do not result in an electric current that is really a good copy of the sound waves. And so manufacturers must always take care in selecting good crystals and in mounting them carefully so that they give the best results possible.

*The electrodynamic microphone* (Figure B). In this

# TRANSDUCING ELEMENTS

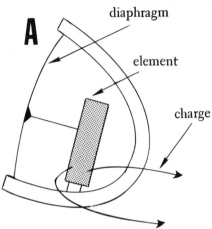

**A** diaphragm

element

charge

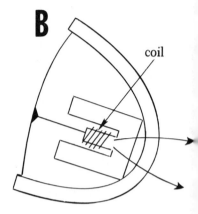

**B**

coil

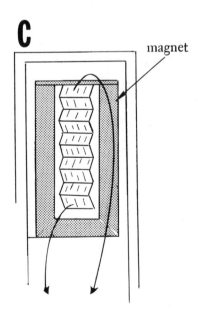

**C**

magnet

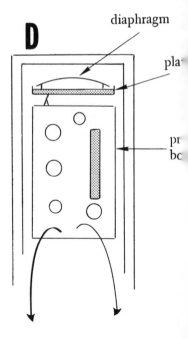

**D**

diaphragm

pla

pr
bo

mike, which is usually referred to simply as a dynamic microphone, the element consists of a wire coiled around one stem in a permanent magnet. When the diaphragm vibrates, it sends the coil sliding back and forth along the stem, actions that create a small voltage in the coil. Born is the electric current.

The dynamic mike is superior to the piezo, generating an electrical wave that is a better copy of the sound wave. At one time, the dynamic was a pretty expensive piece of equipment, but now, thanks to many improvements in manufacturing techniques, it is quite reasonably priced. Some manufacturers include it among the accessories that come with their recorders and decks.

*The ribbon microphone* (Figure C). This microphone takes its name from the accordion-shaped ribbon of aluminum that hangs down through its interior. Suspended between the poles of a powerful magnet, the ribbon serves as both diaphragm and transducing element. It vibrates when struck by sound waves and creates an electrical current within itself. The current passes into attached wires and flows to the recording unit.

The ribbon microphone, which is also known as the velocity microphone, is an excellent one and has long been used in radio stations; in fact, some of today's stations are still using ribbon mikes that were purchased several decades ago. Long-lasting though it is, the ribbon does have the reputation of being a fragile piece of equipment because a good gust of wind can easily break the aluminum strip. If you have a ribbon mike, take care not to use it outdoors on breezy days. And, of course—as with all of your equipment —always handle it gently.

*The condenser microphone* (Figure D). Here we come to what is perhaps the most popular microphone used in recording today. It picks up a wide range of sound frequencies and turns them into excellent electrical copies.

Inside the microphone, the diaphragm forms one plate of an electrical condenser, or capacitor. When struck by sound waves, the diaphragm presses close to its companion plate. Electrical energy is created between the two plates, an energy that alters its strength depending on how close the plates come together. The energy travels into the pre-amplifier circuit board below the plates. The pre-amp then produces a voltage signal that is sent to the recording unit.

Condenser mikes can be either quite inexpensive or very costly. Inexpensive condensers are built into many of today's portable cassette recorders. Simply constructed as they are, the built-ins do a good job with the speaking voice, but usually give a tinny sound to music and song. Very expensive condensers can easily cost several hundred dollars each and are used mainly for making professional recordings.

## PICK-UP PATTERNS

Microphones are typed not only according to their transducing elements. They are also typed according to pick-up pattern—that is, according to the directions from which sounds must come before a mike will accept them.

In all, there are three pick-up patterns: *omni-directional, uni-directional,* and *bi-directional.* They are illustrated in the diagrams on the opposite page.

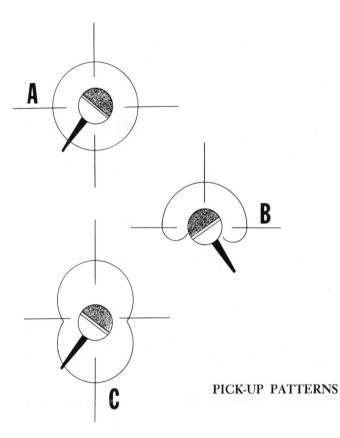

**PICK-UP PATTERNS**

*The omni-directional microphone* (Figure A). *Omni* is the Latin word meaning "all." And so an omni-directional mike is one that picks up sounds arriving from every direction. In a television or a radio studio, you'll often see it hanging from a boom. Aimed downward, it is able to pick up the voices of performers located at different spots on the stage.

*The uni-directional microphone* (Figure B). As the prefix *uni* (Latin, meaning "one") indicates, this micro-

85

phone picks up sound mainly from a single direction. In one way, as can be seen in the illustration, the name really doesn't fit because the pick-up pattern extends out to the sides and can pull in sounds from those directions. But, in other ways, the name is an apt one. First, the mike is "dead" on the side opposite the pick-up pattern and will accept no sounds from there. Second, if you drift beyond 20 degrees to the side of the mike's central axis, your voice will begin to be lost.

If you'll take another look at the uni-directional microphone's pattern, you'll see that it is shaped like a heart. This shape has given the mike a second and widely used name—*cardioid*. The name comes from the Greek word for "heart."

Cardioid microphones have a wide variety of uses. If ever you record a performance in front of an audience, the cardioid will prove especially helpful; you'll be able to pick up the voices and music on stage while eliminating all the unwanted audience noise coming from behind the mike. The cardioid will prove just as helpful when recording in a room that has a high resonance to it; echoes or ringing sounds that would otherwise get into your recording will pretty much go undetected. Finally, you'll find the cardioid a valuable piece of equipment in a public address system; with its confined pattern, it will safeguard you from feedback—the howling noise that deafens everyone when a mike starts to pick up the signals from the public address loudspeakers.

*The bi-directional microphone* (Figure C). This microphone picks up the sounds from two directions—front and back—but cuts off those coming in from either side. *Bi*, of course, is the Latin word meaning "two." Because of its pick-up pattern, it is also known as a *figure-8* microphone.

The bi-directional is especially useful for recording two-way or small-group conversations. Each person or small group takes one side of the mike. No one runs the risk of being "off mike" and going unheard as might happen to someone who gets a bit too far off to the side of a cardioid.

## IMPEDANCE

Microphones can be classified in yet another way—that is, according to what is known as their *impedance*. Impedance means the amount of opposition that the microphone itself exerts against the flow of the electrical current. Other stereo and recording components, among them loudspeakers and amplifiers, also place this opposition in the path of the electricity.

Impedance is measured—*rated* is the word that recordists most often use—in ohms, which are units of electrical resistance. Microphones can be roughly divided into three impedance groups: high, medium, and low. High impedance microphones have impedance ratings up to 10,000 ohms and beyond. Low impedance mikes generally run in the range of 50 to 250 ohms. All mikes between the high and low impedance grades are considered medium types.

If you're to get a good, full sound, your recording unit should be well matched with the impedance rating of your microphone; if the two are mismatched, certain sounds are likely to thin out or be otherwise damaged. There's really no worry here if you're working with a mike that the manu-

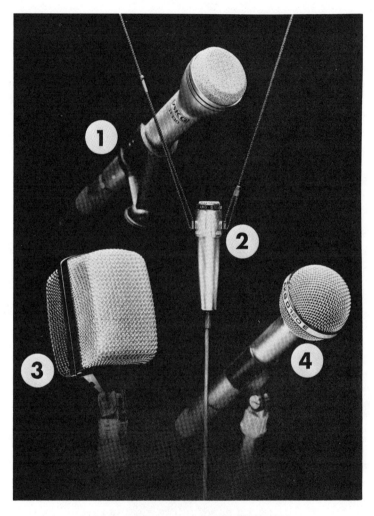

## FOUR POPULAR MICROPHONES

Seen here together are four of today's most widely used microphones. They are (1) the omni-directional, (2) the lavaliere, and (3, 4) two styles of cardioid. (*Photo courtesy Philips Audio Video Systems Corporation*)

facturer sent along with your recording unit; they'll be nicely matched. Further, most of today's recording units can handle impedance ratings as high as 10,000 ohms. However, should you wish to add a mike to your system, make sure that its impedance rating is within the range that can be handled by your unit. A glance at your owner's manual should give you the information you need.

When you begin to record away from home or begin to use greater lengths of cable, you'll need to think more about impedance. For instance, if the cable is the ordinary single conductor shielded type, your mike may begin to lose some treble tones when the cable runs more than fifteen feet in length. You'll also likely pick up some hum in the recording. You may need to go to a heavier cable or a lower impedance microphone. A talk with your local stereo dealer or a fellow recordist should help you solve the problem.

## SPECIAL MICROPHONES

Modern recording needs have caused manufacturers to develop a number of special microphones. These mikes have transducing elements and pick-up patterns that are the same as those in the units we've already mentioned. What makes them special is the uses to which they are put. Take a look at the diagrams on the next page.

*The parabolic microphone* (Figure A). You've probably seen this outfit at sporting events, rock concerts, or political meetings. Designed to pick up sounds coming from a great distance, it is simply an omni-directional mike that is pointed at the very center of a large reflector disc.

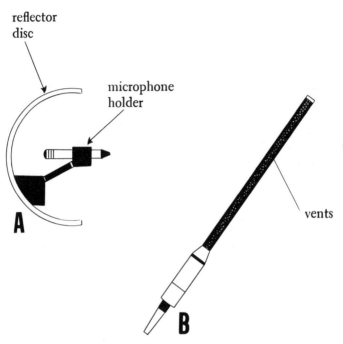

reflector
disc

microphone
holder

vents

**A**

**B**

**SPECIAL MICROPHONES**

Sounds flow into the disc, which is constructed of metal or plastic, and are reflected into the microphone mouth.

The parabolic microphone not only pulls in sounds from a great distance but also rejects much unwanted noise. For this reason, it is especially liked by naturalists who want to record animal sounds while remaining at a distance so as not to disturb their subjects.

*The shotgun microphone* (Figure B). This is a uni-directional mike. Earning its name from its slender, rodlike shape, the shotgun pulls in sounds from a great distance, at the same time cutting out unwanted noise that flows in from the sides. All that you need do is aim it in the direction of your subject.

90

The shotgun eliminates unwanted sounds in a clever way. It is vented all along its sides so that unwanted sounds enter these vents and travel down to the transducing element at the base. The sounds are broken up on entry because some of their components come flowing through vents near the top of the mike, and some through vents near the base. The components then must travel different distances to the transducing element. And so they arrive at different times. Kept from ever "getting together" again as complete sounds, they cancel themselves out and go unrecorded. A clear path is left for the components of the wanted sound. They strike the mouth of the microphone and travel together to the transducing element.

*The lavaliere microphone.* We've all seen this microphone time and again on television. It's simply a very small omni-directional mike that leaves a performer's hands free because it is suspended from the neck or clipped to a shirt or dress. It can also be hidden in the performer's clothing. A lavaliere can be seen in the illustration on page 88.

*The stereo microphone.* Ordinarily, two microphones are required for stereo recording, with each assigned to its own track on the tape. The stereo microphone makes it possible to cover the two tracks with a single mike. Mounted inside its body are two small condenser microphones. They're separated from each other by a distance of about an inch and a half.

*The uni-omni microphone.* This is another two-in-one mike, though it won't produce stereo sound all by itself. What you can do with it is switch its pick-up pattern from cardioid to omni-directional. The cardioid setting can be used for stage work and live music where audience noise is

**TWO PROFESSIONAL
MICROPHONES**
These two professional mikes are the
products of the AKG Acoustics
laboratories. On the left is a condenser microphone that can handle
quadraphonic sound. Its companion
is a two-way cardioid dynamic.
(*Photos courtesy Philips Audio
Video Systems Corporation*)

a possible problem. The omni-directional can take over for
the recording of conversations, interviews, or conferences.

## ALL THOSE EXTRAS

As has been done for all tape and stereo equipment,
manufacturers have turned out some great accessories to
help your microphone do a better job. You'll find a fascinating array of these extras in any stereo catalog or shop. They
include lengths of cable, microphone plugs, mike stands,
booms for suspending the mike at face level or overhead,
and goosenecks. Goosenecks, in case you've never seen
them, are flexible metal tubes that can be attached to mike
stands. Once in place, they enable you to turn or bend the

microphone in any direction with little more than a touch of the hand.

These many accessories can be very helpful. But they can also be on the expensive side and, as a beginner, you needn't rush out and buy them, though you may be tempted to do so when you see them in all their shining glory. Rather, you should work with your beginning equipment, using it to become an expert recordist, and then go looking for a particular accessory at the time you come to need it. In the meantime, however, it will be wise to check the accessories in your local stereo shop and learn about them. Then, when the time comes for obtaining one, you'll be better prepared to make the wisest purchase that your pocketbook will permit.

One accessory needs special mention—the mixer. As you know, all stereo recorders and decks come equipped with at least two microphone inputs so that a mike can be assigned to each track on the tape. Two mikes will serve you well at the start and may be all that you'll ever need. But suppose that, on becoming an expert recordist, you decide to try some work that requires additional microphones. You're going to have to get yourself a mixer.

A mixer is a unit containing the volume controls necessary to bring several sound sources together in a single audio signal. In reality, it's a small version of the giant consoles separated by sound engineers during radio and television broadcasts and recording sessions.

You'll find a wide variety of mixers on the market, with some operating electrically and some electronically. The simplest of them can handle two mikes. Others take a minimum of four microphones, enabling you to use two per track on a stereo tape; if you wish to make a 4-track tape, each mike is then assigned a track of its own. Very expensive units have as many as eight microphone inputs.

Some models also have inputs for a phonograph and a recording unit along with those for the microphones. You can then blend music from a phonograph disc or a tape in with the voices of your performers. You'll be working just as the professionals do in radio and television.

Having become acquainted with the microphone and what it can do, are you ready to give it a try? If so, then let's get to your first live recording. Here we go.

CHAPTER

# MAKING A LIVE RECORDING

A DONOR recording is challenging and fun to make, yes, but a live recording is far more challenging and a lot more fun. When working with a donor, you're concerned mainly with maintaining a proper recording level. Live recording, however, gives you a number of additional requirements to think about.

For instance, there's the problem of helping your friends sound their best on the tape. No matter where and when you make the recording—at a party in your living room or during a class or a stage performance at school— some of the participants are sure to be facing a microphone for the first time. You'll need to show them where to stand at the mike. And you'll probably have to give them some tips on how to speak or sing into it.

Then there's the problem of placing the microphone in the proper spot. Microphones can't be set up in any old

way if you hope to make a good recording. There are recommended set-ups for speakers, for singers, for musical
groups of different sizes, and for a singer who plans to accompany himself or herself on a guitar or a piano.

On top of all else, there's the problem of the acoustics
in the room where you're working. They have a great effect
on a recording, and you may have to alter them somewhat
before you can make a good tape.

In this chapter, we're going to be talking about how
you can handle these various problems. They're basic to all
types of live recording. Once you know how to take care
of them, you'll be ready to set your reels turning, no matter
where you may be.

## LET'S START WITH ACOUSTICS

*Acoustics* is a term that refers to the way a room handles the sounds inside it. Some rooms, especially those with
high ceilings, are said to be acoustically reverberant. Any
sound heard in them tends to resound and echo to one
degree or another, with recordists describing the sound as
full and bright. Other rooms are acoustically dead: Their
furniture, draperies, carpeting, and wall textures take all
the brightness out of a sound and make it seem very dull.

What you're looking for is a room that happily combines these two factors. You want a place that is reverberant
enough to give you a sparkling sound, but dead enough to
avoid heavy echoes. To check the acoustics, have someone
speak to you from across the room, or make a test recording
that runs for two or three minutes. Your ears will tell you
all that you need to know.

A particularly good test is to walk about the room while

clapping your hands together. If there is too much reverberation, you'll produce sharp and ringing sounds. In a dead room, there will be a thudding sound every time your hands come together.

If a room poses any acoustical problems, you can solve them in a number of ways. For instance, you can cut down on reverberation by pulling the draperies closed or by hanging a blanket across a window. Or you can bring in some furniture. Also, if you have a cardioid mike, be sure to use it whenever the room proves too bright; by accepting sound from just one direction, it eliminates much echo. In a dead room, try having the performers stand a little farther away from the microphone than they usually would.

Incidentally, always be sure that the echoing sounds in a live room aren't given a chance to cause vibrations in the stand to which the mike is attached; they'll add to the reverberations being picked up. If the microphone stand is on a table or a bare floor, place a bit of carpeting or padding under it.

Once you've taken care of any acoustical problems, it's time to put your microphones in place. As you know, proper mike placement depends on whether you're planning to record speech, song, or instrumental music. The job of placement for the recording of speech and song is pretty simple, so let's start with it.

## MIKE PLACEMENT: FOR SPEAKING AND SINGING

The microphone should be placed about two feet away from the speaker. If you're using a uni-directional mike, start by aiming the mouthpiece directly at the speaker and then have him or her say a few words while you check the VU meters and set them at a good level for recording. Depending on the strength and quality of the person's voice, you may want to move the individual away from the mike or bring him or her in a little closer. You may also hear that he or she pops *b* and *p* sounds or hisses *s* sounds. These are very common speech peculiarities that can really distort a recording. You can reduce or eliminate them altogether by angling the mouthpiece away from the speaker so that the person is talking *cross mike*.

An omni-directional mike should be placed somewhat below the speaker's mouth level. The voice will then be picked up as it travels across the mike, and any troublesome popping and hissing sounds will be reduced. Or, as you'll see later, the mike can be suspended above the person's head.

Like many people, your speaker may have the habit of leaning forward whenever he or she talks, as if the mike won't have the power to pick up the voice unless the speaker is right on top of it. At the same time, the speaker will probably raise his or her voice. The result: The needles on your carefully set meters will bounce clear across the danger zone and cause enough distortion to last you a lifetime. Take a little time to coach your speaker in the art of always standing straight and talking in a normal one.

There may be a moment in the recording, however, when you want him to assume an intimate tone. Closeness to the mike will give you the needed intimacy, and so he can lean forward at that time. But, to keep the recording level where it belongs, he must also soften his voice a bit. And he must be coached to straighten again when his voice returns to normal.

By the same token, there may be a moment when he is to raise his voice. Coach him to lean far back or take a short step backward. If he forgets, stand by to adjust the gain. He is, of course, to return to his original position once the loud passage ends.

Now let's say that you're working with two people. Place them also about two feet away from the mike and then set your meter levels while each takes a turn talking. If one speaker's voice is much louder than the other's, ask her to soften her tone or step back a little. Or call for the softer voice to raise her tone or move closer to the mike.

Continue testing the speakers until their voices are balanced, meaning that both voices will be recorded at approximately the same meter level. The same job must be done if you have several people clustered about the microphone. Because of the various voice strengths involved, you'll probably need a little more time to get everything in balance.

Let's talk a moment about putting several people in place. If you're using an omni-directional mike, have the participants form a circle around it. Or see if you can hang the microphone about two feet above their heads. This can be done by using a boom stand or suspending the mike from the ceiling. Should you choose the latter course, be sure that the cord used is strong enough to hold the mike (a fishing line is a good choice here) and that it's securely fastened to the ceiling. It's no fun to drop a microphone at any time, especially during a recording.

A uni-directional mike can be a problem for a group. If your equipment includes two cardioids, don't hesitate to use them both, dividing your speakers between them. Should you have just one mike, place your people with particular care; though the cardioid can pick up sounds angling in from the side, be sure that no one is placed so far to the side that his voice will be lost. With an especially large group, some people may have to step aside to make room when it's time for someone else to speak. Have everyone move quietly so that no unwanted noise is picked up— and quickly so that the newcomers to the mike are never late in beginning to speak.

Singers are as easy to place as speakers. When working with a soloist, start with the customary distance of two feet. Then have the singer go through the number (or a major portion of it) once or twice while you locate the proper recording level. At that time, you can adjust the distance as seems wise. It may be necessary for the singer to take a step back from the mike, or at least lean away, when hitting a high or particularly strong note. Of course, your soloist can move in a little closer during softer passages.

Start with the two-foot distance when working with a singing group and again run through the selection once or twice. Listen closely for the strong and the weak voices; move the strong singers to the rear and bring the weaker vocalists to the front. Also, listen carefully to the words being sung. Many groups have the bad habit of garbling things. If you can't make out the lyrics, look for one or two singers with particularly good enunciation and move them to the front. They'll do much to make the words come through in an understandable fashion.

But watch out how far forward you bring them. Should they come too close to the mike, they may well drown out everyone else.

A special point: In setting your levels for a live record-

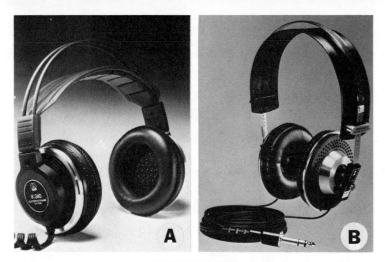

**HEADPHONES**

Headphones are especially helpful when running your tests for a live recording. They can also add greatly to the enjoyment of listening to music. (*Photo A courtesy Philips Audio Video Systems Corporation; Photo B courtesy U.S. Pioneer Electronics Corporation*)

ing, especially a musical one, it's a good idea to wear a set of headphones. In a crowded and perhaps busy room, the headphones will enable you to hear clearly the sounds to be recorded and will give you a true "picture" of what's going to be on the tape. Headphones range in cost from the very inexpensive to the very expensive. A good starter set can usually be purchased for around $20, sometimes for much less if you keep your eyes open for sales at the local stereo shop.

*MIKE PLACEMENT: FOR
INSTRUMENTAL MUSIC*

You'll have to think about a number of things when you begin setting up your microphones to make a recording of instrumental music.

For instance, you'll find yourself concentrating on the different volume of sound produced by the various instruments. These differences can be great and can cause you trouble. If you're not careful, for instance, the powerful brass section in an orchestra can easily drown out the string section. Your mikes must be placed so that you end up with a fully rounded sound in which all of the instruments can be heard.

And you'll find yourself thinking about the kind of music that's being performed. Are you recording your school band playing a march that should have an exciting, resonant sound to it? If so, your mikes should be placed at a fair distance from the band so that they'll help matters by picking up the resonance in the auditorium.

But what if you're recording a small-group selection that will sound better if each instrument is heard more clearly? The mikes will have to be brought in closer.

And what if the musical group is to accompany a singer? You'll need to give the vocalist a separate microphone, plus the instructions to stand in fairly close so as not to be overwhelmed by the accompaniment.

These and many other factors—including the number of musicians in a group and the loudness of their play—will determine where your mikes will need to be placed. Because these factors are bound to change from group to group, each new recording session is going to present a fresh challenge. You'll have to set your mikes in what seem to be good starting positions and then make test runs. Between runs, you'll adjust the positions. At last, when you're satisfied that you've located the best spots, you'll be able to start the recording.

In the beginning, you'll probably have to do far more testing than would be necessary for a voice recording. All your fiddling may tax the patience of the musicians. Work quickly, but tell them—as cheerfully as you can—that you're striving to get the best sound possible. With increasing experience, you'll be able to finish the job in much less time. One day you'll be expert enough to meet that greatest of challenges—the successful recording of a musical performance when there's neither time nor opportunity for even one test run.

When setting up your mikes for the first time, you'll have to use your best judgment as to location, but you needn't work totally by guess. Here are some suggestions for starting placements that can then be adjusted as you go along. Most of them assume that you'll be using at least two microphones.

Let's start with a small band. If you're working with omni-directional mikes, try placing them about six feet in front of the group and five to six feet away from each other. Uni-directionals should be tried about eight feet away from the players and around four feet from each other. For very good results with uni-directionals, aim them slightly inward so that they begin to overlap at the center of the group.

Now suppose you're working with an ensemble, say a trio or a quartet. For a start, set the mikes about five or six feet in front of the musicians. Try a between-mike distance of eight feet. Again, when using uni-directionals, aim them slightly toward the center of the group.

For both the band and the combo, the mikes should be set somewhere between six and ten feet above the floor.

Next, suppose you're recording a soloist at a grand piano. Open the lid of the piano and aim a uni-directional microphone down into the interior. Then place a second uni-directional alongside the keyboard, setting it at about the pianist's face level and angling it downward and slightly toward the interior. Placed in this manner, the mikes should do an excellent job of picking up all treble and bass tones.

Let's say that you're lucky enough to own a ribbon mike. Leaving the piano lid up, again aim your uni-directional into the interior, but place the ribbon mike several feet behind the uni. Set the ribbon at a height of about five or six feet, with one live side aimed at the piano. That live side will pick up the piano sounds directly while the opposite side adds to the beauty of things by catching the room resonance.

If you'll experiment a little by varying the distance of the ribbon mike from the piano, you'll find that you'll achieve different sound colorations. Pick the one that you like best for the actual recording.

What if the pianist is going to sing while playing? First, close the lid of the piano and try working with just one mike. Set it alongside the performer at about face level and aim it along the keyboard. The musician will then be singing cross mike, with less chance of putting any hissing or exploding sounds on the tape. If you'd like to try a second mike, place it a few feet away from the piano and see what sort of an effect you get.

A caution: Never—but never—set the singer's mike on the piano lid. Placed there, it will pick up all the vibrations from the instrument.

So far, we've talked about a grand piano only. But,

sooner or later, you're going to come up against an upright model. If it's by a wall, try moving it out into the room and then aiming your mikes at its back. You should get a nice recording.

Now let's talk about one of today's most popular performers—a singer with a guitar. It's best to work with unidirectional mikes, aiming one across the singer's face (again to take advantage of the cross-mike technique) and pointing the other at the instrument. Start with the mikes about two feet away and then reposition them according to what you hear on your test runs.

Our final example of mike placement takes us clear across the scale, from the soloist to a large orchestra or chorus. Placement here can be ticklish at times. Start by remembering that you want the mikes to cover the entire group. So angle them in a way that will enable them to pick up the sounds from the extreme edges of things, but take particular care not to leave a blank spot in the middle. You'll find that omnis and ribbons (because of the figure-8 pattern) are easier to work with here than uni-directionals. Get the mikes well above the heads of the performers—from six to ten feet above floor level.

At what distance from the group should the mikes be placed? The answer depends on the kinds of mikes being used and on how far across the stage the group stretches. Place the mikes far enough out to the front so that they can be aimed to pick up the sounds on the sides without leaving that blank spot in the middle.

If you have enough microphones for the job, you can do what the professionals do when they record an orchestra —assign a mike to each section. Make sure, however, that you don't point the brass section's mike right at the trumpets; their healthy blare is likely to drown everything else out. With one or two ribbon mikes, you might try another pro technique: Set them eight to ten feet above the orchestra or chorus and then aim one live side down at the performers. The opposite side will take advantage of the room resonance.

## SOME FINAL POINTS

Once your microphones are properly positioned, you'll find the following tips to be of help.

Live singing and music should be recorded at a lower level than music from a phonograph or radio. You'll recall that phonograph discs with popular tunes on them are recorded at a constant volume level and are further helped by the radio station's control of the sound. Live music has none of these protections and can all too easily hit those peaks that bring on distortion.

Further, amateur musicians may not play for the recording in the same manner that they played for the test runs. In the excitement of now performing "for real," they'll probably play much louder and much more enthusiastically than before, driving your needles right off the meters. And so, in all cases, you'll be wise to start with your levels a little on the low side and then, if need be, bring them up as you go along.

Because of the enthusiastic sounds that will come bursting through, you'll need to ride the gain during the recording. Remember to make all level adjustments smoothly.

So that you'll always remain free of those annoying fading-in and fading-out sounds, never twist the controls violently. And, above all, never adjust them unnecessarily just to be doing something.

Finally, if you're recording a stage performance, always fit your machine with more tape than you think you'll need. From the rehearsals and test runs, you may have a good idea of how long the performance will last. But, when the real thing starts, all of the actors, singers, and musicians may take a longer time than expected. And don't forget the audience; their laughter and applause are going to eat up some additional time. You'll go straight up the wall if your tape runs out a few seconds—or, worse, a few minutes—before the performance is over. So have plenty on hand.

CHAPTER

# RECORDING FOR FUN

ONCE YOU'VE made your first tapes, you'll be able to put your recording unit to all sorts of uses. In this chapter you'll find a number of activities that have been enjoyed by count-less recordists. They range from searching for sounds to editing your tapes. Shall we join the fun?

## *SEARCHING FOR SOUNDS*

The world is full of interesting sounds. They're to be heard wherever you go, even if you take but a few steps from your home. As a recordist, you can do far more than just listen to them. You can make a fascinating hobby of putting them on tape.

All that you need do, for instance, is take your unit for

a walk in the out-of-doors. There, singing for your tape, will be some robins. Or there for the tape will be the roar of the ocean as it crashes against a rocky shore. Or the gentle sound of a brook. Or the rustle of a waterfall. And just listen to those crickets at twilight.

If you'd like to do even more, why not take your unit along when the family goes on its next vacation? How about taping the voice of a ranger as he or she leads your group through a national park? And, should you arrive in New Orleans at the right time, how about picking up the varied happy sounds of Mardi Gras? And wouldn't it be great to record the music, applause, and bursting firecrackers that always accompany the Chinese New Year parade in San Francisco?

Wherever you venture, you'll encounter sounds of all kinds, each so different from all the others. To catch them, you're going to need a battery-operated portable unit. It can be either an open-reel or a cassette machine.

As usual, the open-reel machine will give you the best sound possible, but remember the drawback of its cost (at least several hundred dollars). And there's the drawback of its weight. Just about the lightest unit you can buy is going to weigh around ten or eleven pounds and, after a short while, it will become burdensome to carry. The cassette, on the other hand, weighs very little and is quite inexpensive. And, thanks to the manufacturing improvements that were mentioned earlier, it produces a good sound, one that can come quite close to that of the open-reel in quality.

As a beginner, you'll be wise to start with the best cassette that you can find within the limits of your budget. Then, when you're really sold on "sound hunting" as a hobby and have saved enough money, you can invest in an open-reel unit. At that time, your cassette can become a backup unit, to be brought out in cases of emergency.

Incidentally, when you purchase your first portable for outdoor work, make sure it has an adapter that can be plugged into the receptacle for the cigarette lighter in your family car. The attachment can be used when you're recording near the car and will do much to preserve the unit's batteries.

The best all-around microphone for sound hunting is the cardioid. With its uni-directional pick-up pattern, it pulls in the sound you want while pretty well cutting out all the surrounding noise. It's particularly useful on a nature-recording expedition when, for instance, you want to isolate the song of a bird from all the other forest sounds. You can help the mike catch distant sounds by fitting it into a parabolic reflector.

Reflector discs for amateur use are often made of plastic and come in diameters of six or three feet. The three-foot diameter will be best for a start because it can be carted around or stored in the car more easily than its larger brother. You'll find inexpensive parabolic reflectors on display at most stereo shops. They're usually built so that they can be held in the hand or mounted on a tripod. Most have a fitting that enables them to hold microphones of various sizes.

You might also want to think about using a shotgun microphone for your outdoor work. It doesn't reach out to the same distance as a parabolic mike, but it does a better job of picking up bass tones and may be just what you need for certain recordings.

The sounds that you collect may be not only interesting but also valuable. For instance, some might be used as off-stage sound effects in your next school play. Or you might record some new ones especially for the play. Or, as a friend of mine did, you might put a series of your sounds together to earn a good grade in class.

My friend had made a large collection of bird calls. He edited a number of the calls onto a single tape and then inserted a narration about the birds between each call. The tape brought him a high grade in a nature study class. So did a tape that he made on how various dogs sound when they bark; his listeners were surprised at how different the barks were from each other.

## RECORDING FOR PARTIES

A tape unit can do much to make a party a great success. To start, it offers the best way possible to provide continuous music for dancing in someone's home.

### Music

There's always a problem when you dance to music from a radio or a phonograph. With radio music, of course, you have to stop every few minutes and wait around while a commercial or some announcement is aired. So far as a phonograph is concerned, the discs being played often come from albums. An album usually features music of just one kind, and it may not be the sort that everyone present likes or can dance to. And so some of the guests end up sitting over in the corner and growing more bored by the minute. By taping a variety of music before your friends arrive, you can do away with these problems and guarantee a successful get-together.

A tape can also provide uninterrupted background music for dinner parties or for those times when the guests just want to sit around and talk. You can put several hours of background music on a single open-reel tape by running the tape at $1\frac{7}{8}$ ips or $^{15}\!/_{16}$ ips. Yes, I said $1\frac{7}{8}$ or $^{15}\!/_{16}$. Even though both are poor speeds for music, they'll work fine here because you want the music to be heard faintly—just loud enough so you know it's there, but not loud enough to interfere with the conversation. And so you needn't worry so much about the quality of the sound.

## Games and Stunts

It almost goes without saying that a tape unit can be used for games and stunts that will turn a party of any kind into a great success. The guests are sure to be excited and challenged by the prospect of speaking into a mike. And who isn't fascinated by the sound of his or her voice on playback? Here now are some ideas that you might want to try the next time you and your friends get together.

*Candid mike.* This is one of the oldest, if not the oldest, of recording stunts. It's similar to the popular television show, "Candid Camera." A recorder is hidden in one part of the room and then picks up the conversations of the guests who come near it. Later, the recorder is brought out and the tape is played back for everyone present.

Though candid mike can be fun, you have to be very careful with it. Before playback, run the tape through in the privacy of another room to see if it contains any too-personal, offensive, or embarrassing material. If so, erase the objectionable passages immediately or drop the plan to play the tape for all the guests. If you have to erase any passages, be sure to check the tape once again to be sure that you've gotten rid of every last objectionable word.

It's also best to tell the guests that they've been recorded and to obtain their permission for the playback. Some people may feel that their privacy has been invaded, and you run the risk of losing some friends if you don't give them the chance to veto the playback. If any guests don't want their sections played back, go along with their wishes.

You'll find that candid mike works best at small parties made up of friends so close that there are few secrets among them.

If you don't like the idea of hiding the recorder, you can work the stunt in another way. Just place the microphone in full view and run the tape as the guests play a word game such as twenty questions. The playback is sure to be filled with hilarious moments.

*Mixed-up newscast.* The game starts when you seat eight or nine guests in a row and then explain that many people become confused about the facts when they have to repeat something they've heard just once. Now, using a news item from today's newspaper, you want to determine whether the guests share this problem or are blessed with extraordinary memories.

With the tape running, read the news item to the first

guest and then, while you hold the mike, have the guest repeat it in his or her own words to the next guest. For best results, be sure to pick the news item with care. It should be brief (a long story will strike everyone as being unfair), but it should contain several different facts. It will be all the better if it also features a couple of strange or tongue-twisting names. Finally, read the story fairly rapidly to the first guest and set the rule that no one may ask questions or take a look at the news item itself at any time. The players have to depend entirely on their memories.

Move from guest to guest with the mike as the news item travels down the line. By the time you reach the final guest, you should have a completely mixed-up version on your hands—or a completely new story. On playback, it's fun to listen for those moments where the story really went awry as it was being repeated.

*Imagination.* This is a variation of mixed-up newscast. It's particularly good because everyone in the room can join in. Just start things off by making up the opening sentence for a story. Then, as you move about the room with the mike, each guest must develop the story by adding a new sentence. You can count on the story taking some odd turns before you reach the final guest and begin *fast rewind* for the playback.

*Truth or consequences.* This is a very old party game that can be easily adapted for taping. Simply put on a television-type quiz show, asking questions of selected guests. For missed answers, the guests are "punished" with a consequence, meaning that they have to recite a tongue-twisting verse, imitate an animal, sing a song, or attempt to spell some impossible word. Correct answers, on the other hand, earn a prize—always a silly one. Extra fun can be had by having the guests unwrap their prizes while you're "on the air" so that the mike picks up everyone's reactions to whatever comes out of the box.

A tip: Let the quiz show run for no more than ten minutes or so. Like all games, truth or consequences can become tiresome for some guests if you let it continue too long. In fact, it's a good idea to keep all tape games a bit on the short side. If everyone is having fun, you can run on for a bit longer. But, if you detect signs of boredom, finish the game off as neatly and as quickly as possible. You should always remember the saying that professional performers have about audiences: "Always leave them wanting more." It works for games as well as shows.

*Switched questions.* This game calls for you to take several volunteers aside and, out of everyone's hearing, ask them each a question. The questions should be the types that require either an opinion or a description in response. Once you have the answers, slip away to another room, erase your questions, and substitute new ones that make the answers sound ridiculous.

For instance, you might ask a guest to describe a witch in a fairy tale. But, when the tape is played back for everyone later, the question has been changed to "Can you describe your new history teacher?" And out comes the answer: "She has a long purple face and scraggly hair. There's a big wart on the end of her nose. All her teeth are gone. There's hair all over her chin."

*The play's the thing.* This one can be a great success if you have a roomful of aspiring actors on your hands. Perhaps you're giving a party for the members of your school or church drama club. Just gather them about the mike and have them put on a radio play.

The game can take any of several forms. By yourself or with a friend, you can put together a ten- or fifteen-minute playlet in the days before the party and have scripts ready for everyone to read. Or you can invent the plot for a short play, assign parts to the guests, and then have the performers make up the dialogue while they're "on the air." An improvised play of this sort can really be fun in the hands of good actors because of the unexpected twists and turns that they can make the plot take.

Suppose that you're giving a cast party after your club has performed a play at school. You can choose a scene from that play for taping. But have the participants exchange roles. They'll be able to make the switches because actors can always remember what their fellow performers said in a scene. But they won't be letter perfect in their new parts. They'll have to do some ad-libbing, and the results can be really funny.

## Children's Parties

If your family is giving a party for a younger brother or sister, don't hesitate to bring out your unit and put it to work. Young children love to hear themselves on tape. All that you need do is interview them, asking them simple questions about themselves, their homes, their toys and playtime activities. Or you might have them sing, individually or in a group. Or see if you can coax some of the more extroverted ones into imitating animal sounds. To you, this may be pretty much kid stuff, but you'll give the little ones some moments of pleasure and excitement that they'll remember for a long time.

## SOME SPECIAL MIKE TECHNIQUES

In a number of the games and stunts that have been mentioned thus far, you've carried the microphone in your hand and have held it toward the guests when it's time for them to speak.

Hand manipulation of a mike requires the use of a few simple techniques if you're to produce a good tape. First, always follow that prime rule that everyone must follow when making a live recording: Never let yourself get too close to the mike. In the excitement of the moment, it's quite easy to bounce the mike up close to your lips, as if it were an ice cream cone. You'll need to concentrate at all times on holding it from six to twelve inches away. At that distance, you'll always be safe from distortion. Of course, you should adjust the distance a bit when you raise or lower your voice.

Second, again to avoid distortion, never thrust the mike right up to the guest's face when that guest is ready to talk. Also, the mike rushing in like a missile may cause some people to back away, lose their train of thought, or stumble over their words. It's best to stand with the mike an equal distance between you and your guest. Aim it in your direction while you're talking. Then tilt it toward the guest when it's the guest's turn. There's no need, incidentally, to snap it over in your friend's direction; a casual movement of the hand will do nicely.

Some guests are going to grab the mike for support and pull it toward themselves as they talk. Gently extricate yourself and, with a smile, signal for them to step back a bit. And some guests, either out of nervousness or as a joke, may try to yank the mike right out of your grasp. Be as gentle and calm as you can, but don't let them do it.

Watch out for nervous movements on your part that can make their way onto the tape. A friend of mine had the habit of drumming the mike stem with his fingers as he talked. His tapes were riddled with thumping sounds until he changed his ways.

Finally, as you're moving about the room, keep track of the mike cord so that it doesn't become entangled with a piece of furniture or someone's feet, your own included. The cord will be easy to control and watch if you hold it slightly out from your body with your free hand. You should also wrap the cord around the leg of the table on which your recording unit is placed. Then, in the event that anyone trips over the cord, the unit won't be pulled to the floor.

## RECORDING AN INTERVIEW

A moment ago, it was suggested that you interview the guests at a children's party. You'll find that interviews will work well at *any* party, no matter the ages of the guests.

Party interviews are informal ones. The questions are made up on the spur of the moment. They're usually quite simple and are often intended to draw silly answers that, much to the delight of the other guests, put the interviewee on the spot. The greatest fun comes during playback when the guests discover that there's a big difference between how they've always *thought* they sound and how they *actually* sound.

One of the most frequent comments you'll hear is a shocked "But I don't talk like that!" Followed by many an assurance that "Yes, you do."

But what about the formal interview, the interview with, say, a student leader or a city official for the school newspaper or radio station? It's a totally different kind of project. It needs to be approached with great care if it is to be successful.

To do the best job possible, you'll need to keep several thoughts in mind. In fact, there are nine in all. Let's talk about them now.

• *Come to the interview well prepared.* Frame your questions ahead of time and write them down. Only when they're written beforehand will you be able to ask the questions clearly and concisely, without all the *er*'s and *ah*'s that can drive your guests and listeners right up the walls. Written questions will also keep the interview moving in the proper direction and, of course, will protect you from forgetting to cover some very important point.

You should memorize the questions, but you should also have them written out and attached to a clipboard that can be kept at your elbow. Then, if your memory falters at any time, you'll be able to pick up the next question with just a glance.

As a courtesy, you should allow your guests to see the questions ahead of time so that they will have the opportunity to marshal their thoughts or collect any extra data that they might need. Depending on the circumstances, the questions can be shown to the guests a few minutes, a few hours, or even a few days before the interview. Should a guest ask about the questions before they're ready, then you should explain some of the topics you're planning to cover.

• *Do all that you can to put your guests at ease.* They may be unaccustomed to a microphone and become self-conscious when they see it, and so avoid making a big production of setting the thing in place. Make sure that your guests are comfortably seated, and quickly determine the proper recording level by having each guest say a few words. Don't have them go through the old "testing 1-2-3" routine because it's likely to make them feel silly and tense up even more. Set the level as you converse with your guests. If you have to give any instructions about, say, speaking louder, let them come out casually, as part of the conversation.

Finally—and this is the most important point of all—pretend to forget all about the mike once the recording level is set. Your "forgetting" will help the guests to forget its presence. Soon, they'll be talking to you in a relaxed fashion.

• *Start the interview with a few simple introductory questions.* Biographical questions are a good choice because anyone can answer them almost without thinking. These first questions will break the ice and help to relax your guests even more. Then, when you feel that they're ready to go, slip into the actual interview questions. The introductory questions and their answers need take no more than a minute or two.

• *Throughout the interview, ask your questions in a natural, conversational fashion.* And, at all times, be friendly and polite, even if you find that you personally don't care for a guest or realize that you disagree very much with what that guest says. And, even if your questions are argumentative or tough, go on being friendly and polite. Your job is to encourage all guests to talk and air their views. A show of animosity or impoliteness can cause a guest to clam up and ruin the interview.

• *Actually listen to what your guests are saying.* Don't be like the many interviewers who are so interested in what they plan to ask next that they pay no heed to what the subjects are saying at the moment. If you're inattentive, your guests will sense it (your eyes will give you away), and they are likely to turn themselves off and not come up with the information you need.

There's another advantage to paying attention. You'll often hear some unexpected or interesting information that needs to be pursued with questions that you haven't even thought of. These extra questions—you'll have to ad-lib them right on the spot—and their answers may turn out to be the high points of the interview.

• *Don't hurry from question to question.* Let each guest give you a full answer before you move to the next question. You'll help yourself immeasurably here by paying close attention. Only by listening carefully will you be able to tell if the guest has finished an answer and is not just pausing to catch a breath or to collect a thought for the next point he or she wants to make.

• *If you wish, you can insert personal comments on some of the answers or add information from your own experience.* This is all to the good because it helps to turn the interview into a relaxed conversation. But don't comment too much. If you do, you're likely to end up with that unhappiest of results—an interview with yourself rather than with the guests.

  • *Without making a big show of it, keep your eye on the tape so that it won't run out in the middle of an answer.* When you see that there is just a little tape remaining, interrupt the proceedings and quickly, but casually, turn the reel over. You can converse with your guests as you do so. Or perhaps it will be a good idea to take a short break for coffee or a soft drink.

  • *This final suggestion is one that you should follow between interviews.* Take the time to watch professional interviewers on television, performers such as Dinah Shore, Dick Cavett, and Mike Douglas. Study the manner in which they conduct themselves and the methods they use to relax their guests and draw information from them. You'll learn many fine interview techniques that you'll be able to put to good use. And who knows? Maybe you're the world's next Jane Pauley or Johnny Carson.

## EDITING YOUR TAPES

As you become an expert recordist, you may want to begin editing your tapes. The main purpose of editing is to remove unwanted or faulty sections from an otherwise fine recording. You may also remove a section from one part of a tape and transfer it to another part. Or, as my friend did when he put together his presentation of bird calls, you can cut sections from several tapes and join them into a single tape.

If you become a serious recordist, editing can be a very enjoyable and creative activity, one that will enable you to produce tapes that are exactly to your liking. Should you give it a try, you'll find yourself working in one of two ways. Your editing will be done either by *cutting and splicing* or by *dubbing*.

## Cutting and Splicing

To cut and splice means simply to remove sections from a tape and then rejoin it. You'll need a razor blade, an Exacto knife, or a pair of scissors to make the cut. A special adhesive tape, called splicing tape, is used to rejoin the cut ends. It's special because it is manufactured so that it will stick to practically nothing except the shiny side of recording tape.

If you wish to make life a little easier, you can purchase a splicing kit. It contains a cutting tool, a roll of splicing tape, a grease pencil for marking the spots where cuts are to be made, and a small block for holding the tape in place while you work. The kit usually costs less than $10 and is certainly worth the money. As a friend of mine says, you'll need three or four hands to juggle the tape if you try to work without it.

Using the kit, let's try a cut and splice. Begin by running the tape on playback. Stop the tape on reaching the place where the cut is to be made. By hand, wind the tape back past the recording head until you hear the exact spot for the cut. Mark the spot with the grease pencil (taking care not to touch and stain the recording head), remove the tape, and make your cut straight across its width. Now continue on playback until you reach the end of the unwanted passage and then make your second cut.

Next comes the rejoin. At this point, the little block

in the kit comes into play. Running along its length is a narrow channel. Place the cut ends of the tape in this channel. You'll find that it's a shade narrower than the tape and so will hold the ends firmly in place by friction.

Place the ends in the channel so that one slightly overlaps the other. Then, passing your cutting tool along the little groove that bisects the channel at a 45 degree angle, cut through both tape ends. When the residue is brushed away after the cut, the ends will be neatly butted up against each other. Lock them together by pressing a small piece of splicing tape into place across the cut.

A final job remains. Remove your handiwork from the block and carefully trim little arcs into the tape edges alongside the splice. The arcs will narrow the tape into a slight hour-glass shape. This "slenderizing" is necessary to keep the tape from sticking when it passes the heads and roller guides in the future.

And that's it—except for one point. If you're to avoid spilling tape all over the place between cuts, you'll need to work with three reels—the supply and take-up reels, plus an empty spare.

At the start, of course, the tape runs from the supply to the take-up reel. But, after making your first cut, remove the take-up reel and replace it with the spare. Then run the next stretch of tape—it will be the unwanted portion—onto the spare as you move forward to the point where the second cut is to be made. On finishing with the second cut, put the take-up reel back on the machine. You'll be ready for the splice.

This procedure of exchanging the take-up and spare reels should continue for as long as you're working on the tape. By the time you complete the job, all the unwanted portions will be collected on the spare reel. Then, if you wish, you can splice them together, erase the material on them, and use the tape for a recording.

Although the splicing kit is a great help, you'll find a splicing machine to be even more helpful. This is a small hand-held device into which the tape ends are placed. You simply press a level or panel, and the machine does the rest. It cuts through the ends, presses the splicing tape in place, and then trims the tape edges into the hour-glass shape.

The cut-and-splice method of editing, though much used, has at least two shortcomings. First, it can be a real headache when used on tape that's been recorded in both directions. Because you must cut through the entire width of the tape, you're going to sacrifice the material that's been recorded in the opposite direction. And so, if you know that there's an editing job in the offing, be sure to run the tape in one direction only.

Further, splices tend to reduce the strength of a tape. Should you play a spliced tape quite often, you can bet that it will break fairly soon.

### DUBBING

Dubbing requires the use of two recording units because it is the process of transferring the taped material on one machine to a blank tape on the other. Once the two units are tied together, the job is a very simple one.

All that you need do is run the two machines together until you reach the portion to be eliminated from the tape.

Then turn off the machine that's receiving the material, but leave the tape running in the companion unit. On arriving at the opposite end of the unwanted section, switch the receiving unit on again. Repeat the process as often as is necessary. By the time you're done, the tape on the receiving machine will contain all the wanted passages.

Many recordists, if they're lucky enough to own two units, prefer dubbing to splicing because it can be done with such ease. But it does have a drawback. Whenever a tape is duplicated, it loses a bit of its tonal sharpness. And so don't be disappointed if your dubbed tape isn't quite on a par with the original recording.

In this chapter, we've talked about a number of activities that should cause the beginner little or no trouble. Are you game now to try two projects of a more challenging nature? Then let's get right to the next chapter.

CHAPTER

# TWO CHALLENGES

THESE TWO special projects will not only challenge your skills as a recordist; they'll also put your sense of showmanship to a good test. Just what are they? Well, to start, we're going to do a slide projection show with sound. Then we're going to add words and music to your home movies.

## SLIDE PROJECTION SHOWS

If your family owns a 35-millimeter still camera, there's sure to be a large collection of color slides somewhere around the house. Perhaps the pictures are of a vacation or a visit to some historical spot; perhaps they're of a family gathering; perhaps they're of scenic views in your area that struck your family shutterbug as beautiful. Often, when visitors drop by, the screen and projector suddenly appear, and everyone is made to sit through a presentation of these photographic wonders.

The chances are that, even with excellent slides, a showing can become pretty boring in a few minutes. There you sit in the dark while your father or mother, or the whole family, tries to tell the story behind each picture. You know how it all can sound: "This is when we took Aunt Josie to Niagara Falls last year. It would be a nice shot of the falls if she wasn't in the way. We tried to get her to move to one side, but it was no use. The poor dear . . . she's awfully deaf."

And so it goes, on and on, until everyone is starting to fall asleep. But you can change things. Slide projection showings needn't be at all boring, not if you get to work with your tape unit. All that you need to do is go through the slides, assemble the best of them into various programs, and then record narrations to go along with each program. No longer is there a voice (or two or three) stumbling around in the darkened room. Now, as the slides are flashing on the screen, there are crisp, clear explanations coming from the unit. Thanks to your imagination, the audience is watching an interesting *show* rather than a haphazard *showing*.

Producing slide projection shows with taped sound can prove to be a great hobby. Once started, you can do much more than simply add sound to slides that are already around the house. You can dream up ideas for new shows, take the photographs for them, and then finish everything off with a taped narration and background music. And what do you know? Suddenly, you'll find that you've got two hobbies—photography and recording.

Before long, you may discover that your shows are of interest to people outside your home. One recordist earned a good grade in a science class with a slide show about two birds that nested on his bedroom windowsill. Another did a fascinating show on the lives of elderly people in the rest home where she worked as a volunteer. Still another was a hit with the scout troops in his town with his slide shows on how to set up camp and cook in the woods. And many recordists have earned extra money by presenting slide shows on local problems and achievements to clubs and organizations. There's no reason why you can't one day match their accomplishments.

## SYNCHRONIZATION

*Synchronization* is a word that you'll have to remember at all times when you begin to combine sound with any sort of film. It means that the sound must be matched with the picture being seen. Obviously, when working with a motion picture, it can be pretty difficult to match the movement of someone's lips to the words he or she is saying. No problem exists, however, with slides. You need only to make sure that your narration concerns the picture currently up on the screen. You're sure to draw laughs if the recorded voice is talking about the entrance to the hotel where you stayed last summer while a picture of the hotel swimming pool is filling the screen.

You may synchronize sound and slides in either of two ways. First, you can use what is called a synchronizer. This is a small, electronically fitted box that is connected to both the recording unit and the slide projector. Whenever your taped narration is being played, the synchronizer causes the projector to advance from slide to slide by itself. You're able to sit back and enjoy the show with everyone else and not have to worry about changing the slides manually.

Here's how the synchronizer works. As you're making the recording of your narration, you flash the pictures on the screen and keep your eye on them. Whenever you're ready for the next picture, you press the changer button on the projector. A new slide appears and, at the same time, the pressing of the button implants a tone on the tape (sometimes the tone is not audible and is really a pulsing action). Then, during playback, whenever that tone or pulse is picked up, it activates the changer and pops a new picture onto the screen.

Some synchronizers do not work by means of a tone or a pulse. Some cause a little slit to be cut in the tape, with the slit then triggering the picture change. Others depend on silence to do the job; whenever you stop speaking for several seconds—usually from four to eight seconds—the changer is activated.

If your projector is a new one, it may be equipped with a built-in synchronizer. Otherwise, a synchronizer may be purchased separately at a stereo or photo shop. It's usually not a costly item, but you may not be able to afford it just yet and so you'll have to choose the second technique for combining sound with slides. It's a very simple method that can work quite successfully, its main drawback being that the changer on the projector must be operated manually during the show. All that you do when making the recording is ring a little bell, click a metal cricket, or snap your fingers whenever you're ready to move to the next slide. During the show, the sound tells the projectionist that it's time to press the changer button.

## PREPARING THE SLIDE SHOW

It will be easier to talk about the preparation of your slide show if we pretend that you're working with pictures the family has already taken. All that we'll say here can also be applied to slide shows that you dream up and then photograph on your own.

To begin, you should establish a basic idea or theme for your show and then have each of the slides make a contribution to the telling of the story. If you use your imagination, you'll be able to find a theme in almost any collection of slides. For instance, suppose that you and your parents, and the family camera, visited a distant city last year. On reviewing the slides, you keep coming across pictures of buildings and locations that played a part in our country's early history. Lightning doesn't have to strike for you to see that the history behind all these spots will make a fine theme.

With the theme in mind, go through the slides and collect all the ones of the historic spots, dispensing with any and all that have nothing to do with the story you plan to tell (later, if you wish, you can put some of the nonessential shots in, perhaps to add a personal touch about your family or to include a bit of humor). Next, arrange all of the selected pictures into some sort of logical order.

This order will depend much on the theme itself or on some story angle that you want to use. Perhaps you'll want to group the buildings and locations according to their age, putting the earliest ones at the top of the list and moving forward in time from there. Or perhaps several of the spots can be conveniently grouped because they all have to do with a single historical figure or incident. Or perhaps you'll decide to arrange the pictures according to geographical location so that the audience will be taken on a definite tour around the city.

Let's say that you end up with slides of a dozen locations and that you have from four to six shots of each location. Your next job is to put each series of shots into a logical order. You should use here a technique that comes from the motion picture: Start each series with what the moviemakers call an establishing shot, or a long shot. This is a representative picture that introduces the viewer to the location and gives an overall idea of what it looks like. It is then followed by pictures that draw closer and closer until they're showing the various parts of the location in as much detail as possible.

Once all the pictures are properly arranged, it's time to write your narration. The narration should be written out in full. You may be tempted to make a few notes on each picture and then try to ad-lib your way through them at the recording session. Forget it, please. If you attempt to ad-lib, you're almost certain to get mixed up at some point or other, with your voice then falling all over itself. Only with a complete script can you have a hope of doing a letter-perfect recording, one that will sound professional.

Don't let the idea of writing the narration make you nervous. Your script doesn't have to be anything fancy. In fact, you'll be wise to make the explanations of each picture simple, brief, and right to the point. You needn't put in a great deal of description or an army of adjectives. After all, the pictures themselves will be up there on the screen for the audience to see. Let them speak for themselves as much as possible.

Also, you needn't fill every moment with words. There can be a few moments of silence here and there. This quiet time will give the audience a chance to concentrate fully on a picture and enjoy it all the more.

When putting the narration together, start by jotting down a few notes about each picture. They'll keep you from getting lost when the actual writing begins. The notes may also show you that some additional information is needed here and there. If so, a trip to the family encyclopedia or the local library will be in order. Or perhaps your parents can provide a few facts. And, if you've brought some pamphlets or books home from the vacation, don't forget to consult them.

As soon as the narration is written in full, you should read it aloud all the way through, nonstop and with your eye on a clock. Read slowly, just as if you were recording, so that you can learn how long your show will run. There is no set length for a slide show, but many experienced recordists believe that it shouldn't run for more than thirty minutes or so. After that, audience interest is likely to start lagging.

If your narration proves to be too long, don't hesitate to cut it down. You may even find that you have to drop a picture or two. And at the same time, check to see that all the pictures get a fair share of the narration. If the narration for some pictures seems too long or too short when compared with the time given the other pictures, you'll have to make some adjustments.

At this point, you're ready for the recording session. You may read the narration yourself or have a friend with performing ability do it for you. Either way, be sure to rehearse several times before setting the tape in motion. There's nothing like plenty of rehearsal to ensure a smooth performance and reduce the danger of flubs.

If you wish, background music can be played behind the narration. It never fails to give a professional sound to a slide show. The music can be blended with the voice in a mixer or can be played on a nearby phonograph so that it's picked up by the narrator's mike. You should keep the music level down while the narrator is speaking and then bring it up in those moments of silence when new slides are coming into place. Be sure to select the music with care; obviously, it should always match the mood that you're trying to set with the narration and pictures.

If you decide to do the narration yourself, you really should find a friend who will be able to handle the music for you. You're going to have your hands full with the script, and you shouldn't be burdened with the tasks of adjusting the music level or changing records while trying to speak without an error.

It's also a good idea to have a friend take care of the projector and push the slide changer button. Again, you'll be left free to concentrate on the narration. You should give your friend a carbon copy of the script so that he or she can follow it and know exactly when to advance the slides. Or you can arrange to point to your friend whenever it's time to move to the next picture.

One word of caution: As you probably already know, a slide projector can be a pretty noisy thing. For a start, it contains fans that are used to keep the projection lamps cool. And it makes anything from a clicking to a thumping sound whenever it removes the current slide and drops a new one into place. The whir of the fans and the noise of the slides changing can easily be picked up on the narrator's microphone and make their way to the tape. To eliminate the problem, try to set the projector as far away from the microphone as possible. An ideal way is to exile the projector to an adjoining room and place the screen so that the narrator can see the pictures through an open doorway.

When you're ready to record, seat yourself comfortably at a table. There should be ample room for the microphone and your script, room enough to keep the papers from brushing against the mike and creating some unhappy noises. And make sure that you have plenty of light so that the script can always be easily read.

All set? Then roll the tape.

### Presenting Your Slide Show

When the time comes to present the slide show, you'll be wise to pretend that your living room (or classroom) is actually a theater. As you'd do in a theater, take pains to have everything set up beforehand. Nothing bores an audience more than having to while away some long moments while the equipment is dragged out from various closets.

Have the chairs, the projector, the screen, and the recorder all in place well ahead of time. And, also well ahead of time, check the equipment to see that it's in proper working order. Be sure, too, that all cords are out of the way so that no one trips over them. Whenever possible, run them along the walls or tuck them under the rug.

One especially important point: When you place the tape unit and the projector on their table, do not allow their cords to hang straight down to the floor. Rather, to repeat a suggestion made in Chapter Seven, wrap the cords at least once around one of the table legs. Then, should someone trip over the cords, the equipment won't be pulled off the table.

It's a nice idea to have background music playing while the guests are seating themselves. A few minutes of music can be easily placed ahead of the narration on your tape.

Finally, be sure that the projector and recorder are warmed up beforehand. And assign a friend to the job of turning out the room lights on your signal. There will then be no delays at all when your show is ready to start.

## MOTION PICTURES

If you're lucky enough to have motion picture equipment that can handle sound, you're all set to go. With no trouble whatsoever, you can make sound movies of your family, a party, a vacation, or a school project. For some very special fun, you can get your friends together and improvise a movie with dialogue and a simple plot.

But, even if you're without sound-equipped gear, you can still get in on the fun. You can add sound to silent film. Sound can be added to film that you've just shot or to film that was taken many years ago.

### ADDING SOUND TO SILENT FILM

Just as there are two ways of adding sound to slides, so are there two ways of giving silent film a "voice." The first is to have a sound stripe placed on the film.

A sound stripe is a coating of iron oxide that runs in a slender line down one side of the film. Accepting incoming signals and imprinting them as magnetic patterns, it works in exactly the same way as recording tape. In fact, there is just one difference between the two. The oxide on recording tape is placed on a base of polyester or acetate—the tape itself. The sound stripe uses the film as its base.

The stripe is added after the film has been shot, developed, and printed. The film is taken to a photo shop and is sent from there to a lab that is equipped to install the stripe. Back it comes, all ready to accept sound. The cost of having the stripe put in place is usually a little less than five cents per film foot.

At this point, the amateur moviemaker needs to attach a small boxlike device to the projector. The box contains a recording head and some electronic circuitry. All in all, it's a little monophonic recorder that imprints the magnetic pattern on the stripe as the projector is running. The attachment is a fairly inexpensive piece of equipment and can be purchased at almost any shop that sells motion picture gear. Some of today's projectors come with the attachment built in.

The sound is sent to the attachment by means of a microphone. Just as you would when making a slide show, you can record a narration while watching the film on a screen. If you wish, you can blend in music by using a mixer or playing a phonograph near the mike. Sound effects can be fitted in at the appropriate moments. And, should you feel especially ambitious, you can have your family and friends speak some words in synchronization with the movement of their lips on the screen. You can get a good effect this way, even if the "sync" isn't perfect. But use the tactic sparingly, when only a few words are involved. Any mis-matching between the voice and the lip movement becomes very noticeable—and very comic—when things run on too long.

If you make any mistakes along the way, you needn't worry. The sound on the stripe can be erased and you can start over again.

The stripe will give you good sound, but there's that same old problem of cost for the beginner. The attachment, though it's fairly inexpensive, still does cost money. So does the job of installing the stripe, even if you're only paying pennies per foot. But, again, don't be discouraged. Remember, there's a second way to give your silent film a voice.

This second way is quite inexpensive. Just put the sound on your tape unit. And then, as you did for your slide shows, run the unit and the movie projector at the same time. You can produce some interesting motion pictures this way, especially if you use only narration and music. Don't try, however, to have your family and friends talk in time with the movement of their lips. You'll fail every time. It's too tough a job to have the recorder and the projector running in such close unison that the voices and the lip movements will be closely matched.

In fact, you may find it a bit tricky to run the tape and the projector in such close harmony that a new scene never flashes onto the screen before the narration for the last one is completed. The secret here is to get the tape and the projector started in unison. Some practice with the two pieces of equipment should help to take care of things.

And there's another little headache that must be mentioned. Sometimes, thanks to slight fluctuations in their running speeds, the tape and the projector will start to drift away from each other during a showing, and there you are again with the problem of the scenes and the narration starting to mismatch. Fortunately, however, many projectors are equipped with a speed control knob that can slow the film or hurry it up. A bit of manipulation here can often correct the trouble. It's also possible to buy an attachment that will enable you to slow the speed of the recording unit.

The problems of putting sound to silent motion picture film may seem to be many, but they can all be solved with some thought and patient work. And, along with so many other recordists, you're bound to take a great deal of pleasure and pride out of inventing ways to handle the obstacles that loom across your path. So don't hesitate to go ahead and become your neighborhood's Francis Ford Coppola.

## Producing a Sound Film

Let's suppose that you've just shot some silent film and had it printed, or that you've come across some reels of old film, say of last year's vacation. You want to add sound, either with your recording unit or a stripe. In general, your job is going to be much the same as that of producing a narrated slide show. But there are going to be some differences.

First—a similarity—you'll want to find a theme for your movie. But, instead of selecting slides, you'll probably need to edit the film down to the scenes that you need and then splice them together in the desired order. Film editing is similar to tape editing, but if you've never tried it before, perhaps your parents or a friend who is an amateur movie buff can lend a hand. Or the dealer at the local photo shop may be able to give you some advice. And don't forget the local library; on its shelves you'll find a number of books containing helpful directions.

Once the scenes are in their proper order, you must review the film while holding a stopwatch. Closely check and then write down the exact amount of time taken by each scene. You may have to view the film several times before you've timed every scene correctly.

These exact times are a necessity if the narration is to be written so that the words for each scene will fit precisely and not wash over into the next scene. In a slide show, you're able to hold a picture on the screen until the narration for it is finished. Not so in a movie. The film is running all the time. You've got no choice but to keep up with it.

Prepare the narration just as you would for a slide show, starting with notes and then writing the script in full. On finishing each scene, reach for your stopwatch and read what you've written. Read aloud and speak slowly, pretending that you're making the recording. You'll learn very quickly if the narration is too long. If so, you'll have to do some pruning. On the other hand, don't hesitate to add a comment or two if too much time is left over.

Once the entire script is written, rehearse it while viewing the film. You'll probably discover a few spots where some last-minute script cuts or additions are needed. From then on, you'll think that you're doing another slide show. Add music, bringing in a friend to handle it, along with a friend to run the projector. Seat yourself comfortably at the mike after making sure that there's ample room for your script and plenty of light to see by. Then roll the film. You're on your way.

And who knows where you'll go from here? Perhaps one day you'll be making a real movie with your friends performing as actors. And perhaps on a later day you'll be producing movies for television or theaters.

Who can tell?

CHAPTER

# HOW ABOUT A STEREO SET?

IF YOU'RE like most recordists, you're going to want to have your own stereo set one day, with your tape unit serving as an important part of it. And so this chapter has just one purpose: to get you ready for that great day by looking at stereo sets and seeing how they work.

## YOUR STEREO SET—THE BASIC PARTS

A stereo set is a home listening center that provides you with high-fidelity sound. It consists of at least four basic parts. These parts may come together in a single cabinet from the manufacturer or they may be purchased separately. Most serious recordists, for reasons that we'll soon see, prefer to buy them separately. When bought separately, the parts are known as components.

The four basic parts are the tuner, the pre-amplifier, the power amplifier, and a set of loudspeakers. In addition, the stereo set may be equipped with a phonograph or a tape deck, or both. Depending on the wishes of the owner, the deck may be an open-reel, cassette, or 8-track unit. Incidentally, there's nothing to prevent you from having more than one deck (say, an open-reel for one type of recording and a cassette for another) if your wallet can stand the strain.

Everyone knows that loudspeakers receive electrical signals from the set and then, by vibrating, pump them out into the room as sound waves. But the tuner, the pre-amplifier, and the power amplifier can seem pretty mysterious to the beginning stereo fan. Actually, there's nothing mysterious about them at all.

The tuner is the component that picks up radio signals from the air. Some tuners are equipped to capture signals from either AM or FM stations, but most are built to handle both kinds of stations. On receiving a broadcast, the tuner turns it into electrical signals, amplifies them a bit, and starts them on their way to the loudspeakers.

En route, the signals pass through the pre-amplifier, where they are further strengthened. Thanks to the controls on the face of the pre-amp, you can alter the amplitude and the tonal makeup of the signals. When you change their amplitude, you simply adjust their volume, either softening them or making them louder. By changing the tonal makeup, you're able to bring out the treble and bass tones to the degree that pleases your ear and makes the music and voices sound best to you.

The pre-amp doesn't just accept signals from the tuner.

It also welcomes those that are sent from the phonograph and the tape deck, and it enables you to adjust their volume and tonal makeup. In doing its many jobs, the pre-amplifier serves as the switchboard for your stereo set. By flicking the switches on its face, you can select the tuner, the phonograph, or the tape deck for use. Because of its switchboard function, many recordists refer to it as the pre-amplifier-control component.

The power amplifier picks up the signals as they're passed from the pre-amp. It gives them a final boost in strength and dispatches them to the loudspeakers. Its job is to bring the signals up to a power level that's strong enough to activate—or drive, as stereo fans like to say—the loudspeakers. The speakers push the sounds out into the room by vibrating. The signals, on reaching the speakers, must be strong enough to set these vibrations in motion.

If you're like most recordists and prefer to buy the four components separately, you can then assemble your stereo set in any of several ways. But, before we get to this job, let's talk for a moment about why so many people prefer to purchase the components individually rather than all together in a single cabinet or, as it's usually called, a console.

The reasons have to do with the quality of the sound produced, and they begin with the loudspeakers. If the speakers are housed in the same cabinet with the phonograph, their vibrations are going to cause its stylus to shake while moving along the grooves in a record. This shaking, in its turn, can cause feedback and distortion. To offset the problem, then, console manufacturers must equip their cabinets with fairly weak speakers. Genuinely powerful out-

## STEREO SET ARRANGEMENTS

If you buy your stereo components separately, you can still house them together in stylish cabinetry. The speakers in Picture A are shown alongside the receiver and phonograph for display purposes only and would be placed in appropriate parts of the room for stereo listening. (*Photo A courtesy Magnavox Consumer Electronics Corporation; Photos B and C courtesy Akai America, Ltd.*)

145

fits, those capable of a full range of sound, could literally knock the stylus right out of the record grooves.

Also, for truly good stereo sound, the loudspeakers should be placed apart and at spots in the room that prove best for them. This sort of placement isn't possible, of course, when the speakers are locked in a cabinet.

Stereo buffs also come up with an economic argument. They point out that, when you buy a console, a good part of your money is spent on the cabinet itself (and, as a visit to a stereo store will prove, the cabinets can be pretty handsome and fancy things). It's far wiser, the fans argue, to invest your money in better components rather than in cabinetry.

Finally, there is the argument that you'll be unable to improve your set if you buy a console. Suppose that you want a better tuner or a better tape deck. You won't be able to make the improvement without buying an entire new console. But, if your set is built of separate components, you can improve one component or another as you go along. Or you can start with a component missing—say, the phonograph—and then introduce it at a later date when you've saved enough money for the purchase.

## PUTTING YOUR STEREO SET TOGETHER

There are four ways in which you can put your stereo set together if you buy the components separately. The method that you choose will depend on what you want your set to do and on the amount of money that you have to spend.

## The Integrated Receiver

Let's start with what is today's most popular approach. Here, you use an integrated receiver or, as it's more commonly and simply called, a receiver. This is a unit that contains, all in one case, a tuner, a pre-amplifier, and a power amplifier. As shown in the illustration, you do no more than connect your phonograph, tape deck, and loudspeakers to it.

### THE INTEGRATED RECEIVER SYSTEM

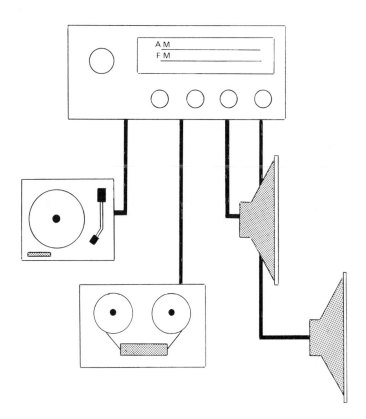

Integrated receivers have come into widespread use in the past twenty to twenty-five years. Before that, it wasn't considered wise to invest in them because vacuum tubes were used in stereo components. The tubes produced a great deal of heat. If the receivers were to generate any real power, the tubes had to be especially large, so large that they produced enough heat to damage the workings of the receiver. And so, for safety's sake, the tuner, pre-amp, and power amplifier were kept separate, preventing the heat produced in each from reaching the others.

But a change came in the late 1950s and early 1960s. Stereo manufacturers began turning to the use of transistors —solid-state circuitry—in their components. Transistors produce heat, it is true, but not to the same damaging extent as vacuum tubes. Reliable receivers came on the market and grew in popularity until today they are found in more than 85 percent of the stereo sets used in the United States.

Their popularity is due to several factors. First, there's the matter of expense. Though the receiver contains three components in one, it costs far less than do the tuner, the pre-amp, and the power amp when purchased separately. Receivers run from under $200 to around $750. A good "starter" receiver can be had for under $200.

Second, there's the matter of convenience. With everything in just one case, the receiver takes up just a little space, a great advantage if your stereo is meant for your bedroom or a small apartment. And, if you're someone who is mystified by the simplest of electrical problems, a receiver doesn't require all the wiring needed to link three components together. Finally, there's no duplication of controls, as there's bound to be when the three components are used.

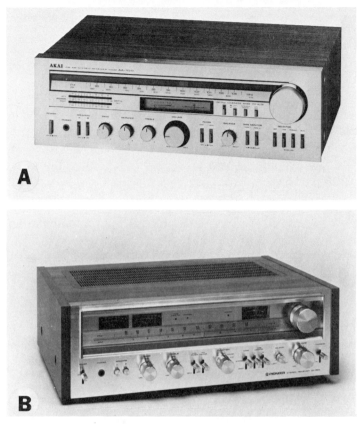

**INTEGRATED RECEIVERS**

*(Photo A courtesy Akai America, Ltd.; Photo B courtesy U.S. Pioneer Electronics Corporation)*

Incidentally, many beginners make the mistake of thinking that the receiver is a radio. A receiver does contain the tuner for picking up AM and FM broadcasts, yes, but it is not technically a radio. A radio contains its own built-in loudspeaker.

THE INTEGRATED AMPLIFIER

If you wish, you can build your set around an integrated amplifier. This is a unit that contains the pre-amp and the power amp. The tuner must be purchased sepa-

149

rately and then tied into it, along with the phonograph, the tape deck, and the loudspeakers.

On first hearing the integrated amplifier, many beginners immediately ask, "Why bother to buy the thing when you then must get a tuner? Why not just go for a receiver instead?"

To answer the question, pretend for a moment that

## THE INTEGRATED AMPLIFIER SYSTEM

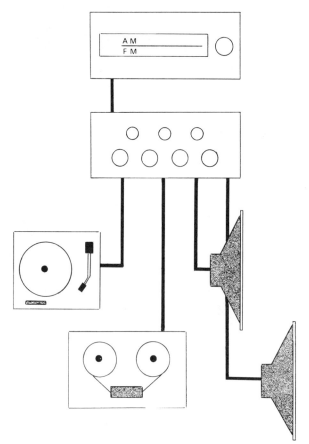

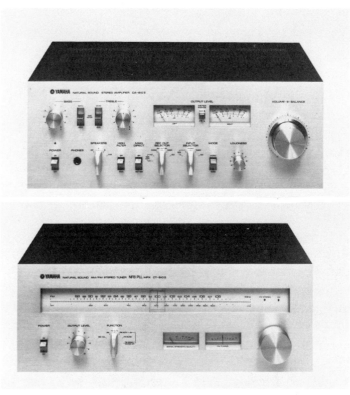

**INTEGRATED AMPLIFIER AND TUNER**
Seen together here are the two basic components in the integrated amplifier system. Shown at the top is the integrated amplifier itself. Below is the tuner. (*Photos courtesy Yamaha International Corporation*)

you're someone who is mainly interested in listening to tapes and phonograph records. AM and FM broadcasts don't mean much to you, and so you have little use for a tuner. With limited funds, you may decide that you'll get the best listening by investing most of your money in a high-quality integrated amp that produces greater power than a receiver in a similar price range. You can then select an inexpensive tuner for what little radio listening you'll be doing, or you can do without the tuner until you can afford it.

The question can be answered in another way. Integrated amplifiers often have controls and features not found in similarly priced receivers. These extras appeal to many stereo buffs, because they help to produce sounds superior to those from the receiver.

## The Three-Component System

Just as its name indicates, the three-component system calls for the use of a separate tuner, pre-amp, and power amp. The tuner, phonograph, and tape deck are tied into the pre-amp. In its turn, the pre-amp is wired to the power amplifier, which then extends its lines to the loudspeakers.

Used by everyone in pre-transistor days, this system continues to be the choice of a few stereo enthusiasts today. It's especially popular with people who enjoy their music at very loud listening levels. The system provides them with more power than the integrated amplifier or the receiver can produce.

The three-component system is also popular among listeners who intend to install loudspeakers in several rooms so that there will be music throughout the home. Additional power is needed to drive the extra speakers. Most receivers and integrated amps have enough muscle to drive from two to four speakers, but, if you're planning to go beyond that number, you're going to need the power of the three-component system.

## The Compact Stereo

Here is a stereo system that has been winning many supporters in recent years. It's similar to a console except that it's much smaller and far less expensive. Contained in

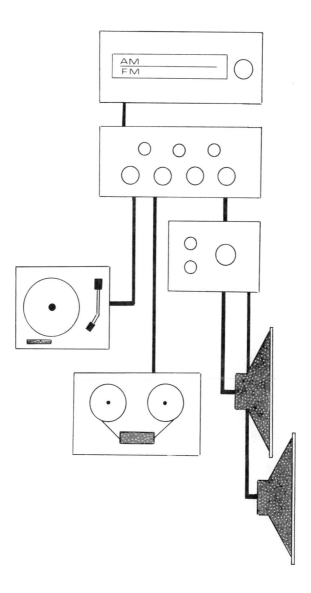

**THE THREE-COMPONENT SYSTEM**

a single unit are an AM/FM receiver, a phonograph, and a cassette tape deck. (For a few dollars less, you can buy a compact that sports just a phonograph or a deck.) The outfit comes with speakers that can be detached and placed a distance from it.

Until recently, most stereo enthusiasts scorned the compact. It was obviously designed, they said, for customers with little to spend. Consequently, its costs had to be kept at a minimum and its manufacturers were forced to cut many a corner so far as quality was concerned. The result was a set that generated very little power and produced below-par sound.

But, in recent years, there has been a growing demand for modestly priced stereos of good quality that don't take up a great deal of space. Among the most eager customers have been students in dorms, people who live in apartments or condominiums, and families in need of a second set for a summer home or a playroom.

The demand has been such that it could not be ignored. Manufacturers have taken seriously to the job of improving the technology in their compacts. As a result, many fine compacts have been appearing on the market.

Compacts are to be found in a wide price range, from little more than $100 to $750 or more. The least expensive sets feature a minimum of controls (usually just enough for basic tuning) and do not put out a great deal of power, but their strength is ample for small-room listening. Expensive models generate plenty of power and can sport all of the gadgetry found on component rigs.

Though a compact can be a good buy for someone with little space, it does have disadvantages. They're the same

**COMPACT STEREO SET**

The compact is winning increasing popularity among people who enjoy stereo music but have little space for the components. The model seen above can receive AM/FM broadcasts and is equipped with a record changer and cassette deck. (*Photo courtesy Zenith Radio Corporation*) as those found in the console. For one, with everything welded together in a single unit, you won't be able to upgrade a component when the time comes; you'll need an entire new set.

For another, again because all the gear is in one box, you'll have to surrender the whole set to the repairman should one part—say, the phonograph—go sour on you. With separate components, of course, you can pull out the malfunctioning part and continue to enjoy the rest of the set.

So far as upgrading is concerned, there are two items of compact gear that can be improved upon at any time— the loudspeakers. This is because they are detachable from the body of the set. In fact, many stereo shops will demonstrate several speaker grades at the time of purchase and then let you take your pick. Naturally, better speakers are going to cost you some additional dollars.

This talk of loudspeakers brings us to one of the most important features in a stereo set, no matter whether it is a compact or a set built of individual components.


## LOUDSPEAKERS

As the instrument responsible for converting electrical signals into sound waves, the loudspeaker has an importance that cannot be stressed too much. If it does a good job, the emerging sounds will be full and rich. But, if it performs poorly, the sounds won't be pleasing to the ear, no matter how fine and expensive the rest of the components in your set may be.

A loudspeaker has two assignments. First, on receiving a signal from the power amp, it must begin the conversion process by setting the air around it in motion, a job that it does by vibrating. Then it must project—send—that motion out into the room in waves so that it can be heard. The speaker is able to do its work because it is divided into two basic parts: a driver and an enclosure. Let's look at each in turn.

## THE DRIVER

Just as the name indicates, the driver "drives" the speaker, causing it to vibrate and set the air in motion. There are several different kinds of drivers in use today. By far the most popular one of all is the dynamic driver. Any loudspeaker that is equipped with it is known as a dynamic speaker.

To understand how the dynamic driver works, we have to look first at the speaker's diaphragm, or cone. Anyone who has ever seen a loudspeaker will recognize the diaphragm immediately. It's the disclike device that's made of stiff paper or a similar material. Circling it and holding it taut is a metal frame, called the basket. The driver itself, as can be seen in the illustration on the next page, is connected to the rear of the diaphragm. It consists of a magnet, voice coils, and wires leading to the voice coils from input terminals.

When signals from the amplifier reach the input terminals, a magnetic field springs up around the magnet. The voice coils are stimulated by that magnetic field and begin to move. Because they are connected to the diaphragm, they cause it to vibrate, with those vibrations setting the

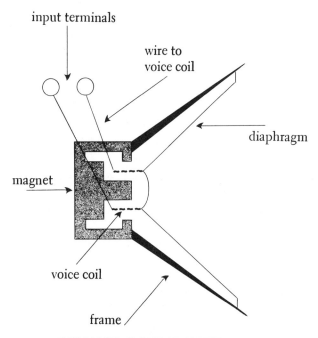

input terminals

wire to
voice coil

diaphragm

magnet

voice coil

frame

**DYNAMIC LOUDSPEAKER**

air in motion and pushing it out into the room in waves.

Though the dynamic driver is the most popular one of all, there are several other drivers that are coming into increasing use. For instance, you'll find some speakers equipped with electrostatic drivers; instead of creating a magnetic field, they generate electrostatic energy—the type of energy that makes your hair stand on end when you comb it on a warm, dry day. A few speakers are powered by induction drivers; here, the diaphragm is ribbed with strips of metal that begin to vibrate when an electrical current strikes them. And some are powered by ribbon drivers; they employ voice coils that trigger vibrations in a metal ribbon, with the vibrations then being funneled into the room by means of a hornlike device.

The variety of signals coming from the amplifier can

cause trouble for any driver. Responsible for the problem is the fact that the various signal frequencies all have different traits. For example, low frequency signals are much more difficult to push through the air than are high frequency ones. This is because there are wide spaces between the low waves and because the waves tend to spread out over a broad area as they travel.

Faced with such problems, sound engineers learned early that a loudspeaker needed to be equipped with at least two diaphragms if it was to give high-quality sound—one diaphragm for the low frequency signals, and one for the high frequencies. Today, we know these two diaphragms by the rather comical names of *woofer* and *tweeter*.

**SPEAKER UNIT SHOWING WOOFER AND TWEETER**
(*Photo courtesy Yamaha International Corporation*)

The woofer is a large, deep-coned diaphragm. Its job is to send the low frequency sounds—all of the basses—out into the room. The tweeter is a small diaphragm that, featuring a shallow cone, handles the high frequencies—the trebles. Each of the diaphragms has its own driver. Speakers containing a woofer and a tweeter are known as two-way units.

More expensive speakers have three diaphragm/drivers —a woofer, a tweeter, and a driver for taking care of frequencies in the middle range. Units such as these are called three-way speakers. If you have enough money, you can even buy four-way, five-way, and six-way speakers. All of them are equipped with additional drivers, mostly for the middle-range frequencies.

### The Enclosure

The enclosure is the case, or box, in which the loudspeaker is contained. As you can see from the photos on the opposite page, it comes in a variety of shapes and sizes. Regardless of its design or size, it has but two jobs—to keep the emerging sounds from escaping in unwanted directions and to aim them strongly in the direction of the listener.

To do its job efficiently, the enclosure must be well built and must have a solid back panel. Always shy away from buying a speaker that has a flimsy cardboard back panel or that, worse yet, has no backing at all. Too many sounds will flee to the rear and reduce the richness of the tones reaching your ears. Also, if the enclosure is cheaply built, you can bet that the speaker hidden within it is of a quality just as poor.

## MODERN LOUDSPEAKERS

From the drawing boards of today's designers are coming some very sleek and futuristic speakers. Some of the best examples are these models from Epicure. The pictures at the far left and right are of the same speaker. At the left, it is seen without its decorative covers. (*Photos courtesy Epicure Products, Inc.*)

### MINIATURE LOUDSPEAKER

Thanks to modern technology, speakers no longer need be large to produce full and rich sounds, as witness this model that's no taller than a book. (*Photo courtesy Bang & Olufsen of America, Inc.*)

161

## *THE PHONOGRAPH*

The phonograph unit in a stereo set consists of five basic parts. Of their number, the largest are the turntable, the tone arm, and the electric motor that gives the unit life. The remaining two parts—the cartridge and the stylus—are attached to the tone arm.

To almost everyone in this day and age, the workings of a phonograph are anything but a mystery. Concealed in the body of the unit, the electric motor activates the turntable and causes it to spin at a steady rate of speed. Also activated by the motor is the tone arm, which tracks its way across the face of the record while the stylus (or, as it used to be called, the needle) travels from groove to groove in the disc. The stylus vibrates against the walls of each groove, with the cartridge picking up those vibrations and turning them into electrical signals that head for the receiver or the pre-amp, their first stop on the journey to the speakers.

### The Motor and Turntable

Depending much on the cost of the phonograph, the motor may drive the turntable electronically or by means of belts and pulleys. So that today's discs can be accommodated, most motors spin their tables at two speeds—45 and 33⅓ revolutions per minute (rpm). Some are capable of a third speed of 78 rpm; that's the speed needed for the records of yesteryear.

Turntables are finely engineered to revolve not only steadily but also at a very accurate rate of speed so that the music will be neither too hurried nor too slow. But, because they're man-made mechanisms, turntables aren't perfect creations and, for one reason or another, their speeds may vary slightly. No quality turntable, however, should ever vary more than 2 percent in accuracy, assuring you good listening at all times.

Many turntables are equipped with adjustment controls for resetting the proper speed when it has been lost. Along with using your ears, you can make a check of a turntable's accuracy by glancing at what are called *stroboscopes*. These are tiny squares that are imprinted all along the bases of many turntables. When the table is not spinning as it should, they'll all run together in a whitish stream. But, if all is well, you'll see the squares clearly, and they'll appear to be standing still.

## THE TONE ARM

The job of the tone arm is simply to carry the cartridge and the stylus horizontally across the face of the record. To do its job, the tone arm must also be able to move vertically with equal ease. This is because the stylus not only vibrates from side to side but also up and down as it travels. Too, many records become warped in time, and the tone arm is then forced to climb over the undulations in the disc.

The tone arm consists of the arm itself and the base that anchors the whole mechanism to the phonograph. The arm is usually built of a strong but lightweight metal. At its base end, it is often fitted with a counterbalance. This is a weight that, when properly adjusted, allows the arm to ride with maximum lightness so that the stylus won't bear down and damage the record or cause distortion.

Many tone arms today come equipped with an anti-skating control. As the stylus approaches the center of a record, it is pulled inward with increasing force because of the narrowing circumference of the grooves. The anti-skating control applies an opposing pressure that keeps the stylus from surrendering to the urge to slide, or skate, across the remaining grooves to the record's nameplate.

## The Cartridge and Stylus

The cartridge is a small housing that is fitted onto the outer tip of the arm and that, in turn, has the stylus fitted into it. There are several different types of cartridges on the market today, but they all generate their electrical currents in the same way—by means of magnetism.

In one especially popular cartridge, the stylus is linked to a magnet that moves freely within an arrangement of coils. When the stylus vibrates, it sets the magnet and the surrounding magnetic field in motion. The movement of the field generates a current in the coils. The cartridge is known as the *moving magnet* type.

Another cartridge features a fixed magnet surrounded by coils. The stylus here is linked to metal strips that are set to moving within the field of the magnet, producing a current that is picked up by the coils. This cartridge is identified as the *variable reluctance* type.

An opposite approach is used in a third cartridge. Now the stylus is connected to two coils. Nearby is a fixed magnet. Vibrated by the stylus, the coils generate a current by quivering within the field of the magnet. The cartridge is aptly named the *moving coil* type.

As for the stylus itself, it may be a needle-and-arm arrangement that is placed in a locking device in the cartridge. Or it may come in a little housing of its own that is then inserted into the cartridge. Almost every stylus is detachable from the cartridge and can be removed for repair or replacement. In a few phonographs, especially those on the inexpensive side, you'll find the stylus to be a permanent part of the cartridge.

In pre-stereo days, the stylus was conical in shape. That shape is still employed in some less expensive styli today, but it really isn't suitable for stereo discs because it doesn't fit neatly into their specially contoured grooves. Most styli are now elliptical in shape when seen in cross section, while others resemble a figure 8 in cross section. In each case, the stylus is able to cling to the sides of the grooves and catch all the little changes in contour that end in strong signals leaving the phonograph.

Fine quality styli are either made of diamonds or tipped with diamonds. The diamond is a strong mineral that will last a long while with proper care, but it will eventually wear out. You'll need to buy a new stylus periodically and so, when choosing your first phonograph, you should make sure that the stylus is the detachable type. Otherwise, you're going to have to replace the entire cartridge.

### RECORD CHANGING: MANUAL AND AUTOMATIC

Over the years, manufacturers have learned that not all people like to operate their phonographs in the same way. Some prefer to handle the records one at a time, putting each on the turntable and then taking it off before re-placing it with another. But others simply want to stack a number of records on the machine and then let it do the rest. To satisfy these opposing tastes, modern phonographs come in three varieties: manual, semiautomatic, and fully automatic.

If you belong to the first group of people, then the manual unit is for you. It must be turned on and off by hand. The discs must be placed and removed by hand. The tone arm must be set in the opening grooves by hand and then lifted away when the music ends. The manual unit is always the choice of that person who enjoys being very much involved with the records, right down to taking plea-sure in their touch. The manual is also often chosen over the automatic unit by the enthusiast who plays classical records, some of which must be turned over before a selec-tion can be heard in its entirety.

If you enjoy handling records but would like your phonograph to have a few convenient features, you'll prob-ably be happiest with a semiautomatic unit. In some mod-els, the tone arm must be placed by hand, but the machine takes over at the end of the music, returning the tone arm to its cradle and then switching itself off. In other models, you need do no more than press a button once the disc has been set in place. From then on, the machine handles all the tone arm actions. All semiautomatics call for you to place and remove each record by hand.

Finally, you may be someone who wants to turn the phonograph on and then forget about it. The fully auto-matic unit was made for you. One after the other, it plays up to six records that have been stacked on the turntable

## MODERN TURNTABLES

Clean lines are the hallmark of the most modern turntables. When purchased separately, all phonograph units come with a plastic lid to protect them from dust. The lids are detachable. (*Photo A courtesy Philips Audio Video Systems Corporation; Photo B courtesy Bang & Olufsen of America, Inc.*)

spindle. Once the unit is turned on, your only job is to listen. The unit drops each record into place, sets the tone arm in position for the start of each, and lifts it away at the end of each. When the final record has ended, the machine brings the tone arm to rest in its cradle and turns itself off.

Manual and semiautomatic phonographs are usually referred to as turntables, whereas the fully automatic unit is known as a changer. Many stereo buffs argue that the changer is not a good choice for those who are serious about the upkeep of their records; they point out that there can be groove damage whenever a number of discs are stacked together. It is true that there is a risk of damage, but it's also true that the risk today is not great. This is because modern discs are made of tough plastic and are manufactured with their centers and outer rims slightly raised so that the grooves are protected from contact with adjoining records while stacked.

**A MODERN CHANGER**
The changer is a fully automatic unit that can play a series of records without stopping. The record holder is seen at the upper left of the unit. (*Photo courtesy Zenith Radio Corporation*)

It is also believed by some enthusiasts that the changer is especially vulnerable to breakdown because, thanks to all its automatic features, it contains a greater number of moving parts than do the manual and semiautomatic units. This may have been a valid argument in the past, but manufacturing improvements have knocked the pins out from under it, except in cheaply built models. A quality changer today is a strong piece of machinery and should match its two relatives in trouble-free performance.

Hopefully, all the facts that you've come across in the past few pages will serve you well when it's time to purchase your first stereo set. But one thing must be said about them. Because the purpose of the chapter has been simply to introduce you to the set, they've all been very basic facts. There is much, much more that you're going to need to know about the various components before you can really call yourself a stereo expert. The components are all fascinating and complex mechanisms, and many a happy year can be spent studying them. If you'd like to begin that study soon, be sure to check the Selected Reading section at the end of the book. The books listed there should be of great help.

And now, in the final chapter, let's turn to a tape unit that records pictures as well as sound.

CHAPTER

# THE MARVELOUS VTR

THE YEAR 1956 was a memorable one for television broadcasting. It was the year that Ampex Corporation introduced
the videotape recorder (VTR) and put it to work. Instantly,
viewers all across the country no longer had to squint at
the fuzzy kinescope recordings of their favorite TV programs. Now there were pictures so clear that they seemed
to be coming from a studio right next door.

In the quarter-century since that memorable year, the
VTR has been developed into a machine capable of doing
many jobs. It no longer limits itself just to recording shows
in a TV studio. It's now taken outside for on-the-spot cov-
erage of news, sports, political, and entertainment events.
It's now found in the home, where it is used to record programs coming from an ordinary television set and where,
thanks to the development of a small camera, it enables

families to put on their own programs. And it now serves a wide variety of instructional purposes in our schools and in other organizations, including industry.

Altogether, the VTR has become one of our most versatile machines. And, considering the many jobs that we may yet dream up for it, it promises to become even more versatile in the future.

## THE PROFESSIONAL VTR

Whether used professionally or in the home, the VTR is an all-electric unit that, though far more complicated, operates just as the tape recorder and deck do. It imprints a combination of video and audio signals on a passing strip of magnetic tape. The tape used for professional recordings is two inches wide. For home use, the tape width measures from half an inch to an inch.

The work of the professional unit begins when cameras convert the images of performers and scenery into electrical signals, while the microphones pick up the audio signals. Everything is sent to the unit, with the video signals (those from the camera) going to one recording head and the audio signals to another.

The video head imprints its signals along a wide band in the middle of the tape. The audio signals land on a narrow track at the top edge of the tape. At the same time, a separate audio signal is planted on a track at the bottom edge. It is needed because, for reasons that we'll see in a moment, the video head spins during recording and playback as the tape moves past. The head and the tape, however, do not run at the same speed, and the control signal is used on playback to keep the head aligned exactly with the patterns on the passing tape.

The developers of the VTR had to overcome an especially great difficulty before their machine could be put to work in broadcasting. To make a good recording of visual objects, the VTR must be able to accept signals of frequencies up to 6 MHz (6 million cycles per second). High as they are, these signals can be recorded if the tape is traveling fast enough. But it must move along at more than 1,000 ips to get the job done. Obviously, promising to eat up some 4 million inches of tape during the recording of a 60-minute program, the speed is an impossible one for practical purposes.

The developers solved the problem by turning things around. Instead of letting the tape whiz by, they made the recording head do the work. They fashioned the head into a disc that spins at a high speed while the tape moves along at about 12½ to 15 inches per second.

Actually, as shown in Figure A opposite, the disc is a four-in-one device. Housed along its circumference at precise intervals are four recording heads. As the disc spins, each head takes a turn implanting its magnetic pattern on the tape. They're spaced so that, as one head slips off the bottom edge of the tape after making a pass, the one following it is already traveling across the video track. Working in this manner, the heads leave no blank spots on the tape.

The tape itself is also made to play a part in avoiding blank spots. By means of a specially shaped block (Figure

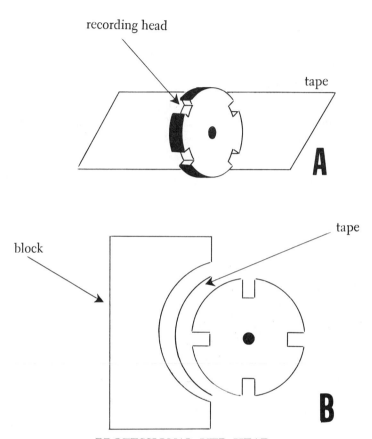

recording head

tape

**A**

block

tape

**B**

**PROFESSIONAL VTR HEAD**

B) that is located directly opposite the disc, the tape is bent into a slight arc across its width. The bend permits the heads to lay their patterns over a wider area so that nothing is lost at the top and bottom of the track.

Because both the disc and the tape are moving, the

173

magnetic patterns are printed as slanting lines across the width of the tape. The across-the-width patterns enable the VTR to store far more material on a given stretch of tape than can an audio unit, which implants the patterns in a horizontal line.

With its four heads, the disc is known as a *quadrature* or *quadraplex head,* names that were abbreviated long ago to *quad head.* It made the VTR a practical television tool. It continues to work in professional broadcasting to this day.


## THE VTR IN THE HOME

Early VTR manufacturers were quick to sense that, if built for home use, their product would offer such exciting entertainment possibilities for the family that it would soon rival the audio recorder and deck in popularity. The professional unit, of course, was out of the question for the home. Its size was too large, and its workings too complex. And, above all else, there was its price. At the time, basic professional units ran from $50,000 to $80,000.

By 1960, however, the Toshiba company in Japan had come up with a more compact and economical recording system that made home units a distinct possibility. Called the *helical recording system,* or more simply the helical scan, it employed a head less complicated than the disc, and a tape that was only half as wide as professional tape.

Instead of the spinning disc, the system is built around a stationary drum. Circling the drum at its middle is a narrow slot. Within the drum, and right in line with the slot,

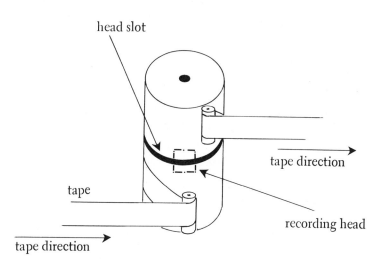

head slot

tape direction

tape

recording head

tape direction

**THE HELICAL SCAN SYSTEM**

is a head unit consisting of one or two heads. As soon as the VTR is turned on, the head unit spins, and the magnetic flow passes through the slot and onto the tape.

As for the tape, it is wrapped in a spiraling pattern around the drum. *Helical* means spiral in shape, and it is this pattern that gives the system its name. On leaving the supply reel, the tape makes a U-turn around a guide roller at the base of the drum. Here, it passes the audio and control-track heads. Then it snakes upward past the slot, makes a U-turn around a guide roller near the top of the drum, and then heads for the take-up reel.

Like the professional head, the spiral enables a great deal of material to be implanted within a given length by slanting the magnetic patterns across the width of the tape.

In manufacture, the tape can be made to spiral in either of two ways. First, it may be spiraled in the *omega*

175

*wrap,* so called because, when seen from the top, it resembles an omega, the twenty-fourth letter of the Greek alphabet. It's a fine wrap that's found in many home VTRs, and it's the one that was used in the above explanation, but it does present a problem. The wrap fails to bring the tape completely around the drum, and so the heads always miss a portion of the picture being recorded. In playback, this loss appears as a narrow band of white at the bottom of the picture.

The loss is not bothersome because it's hidden by the frame around the TV screen and can't be seen unless the horizontal adjustment goes awry and the picture rolls over. Minor though the problem is, some manufacturers prefer to sidestep it altogether and use what is known as the *alpha wrap.* Named for and resembling somewhat the first letter of the Greek alphabet—alpha—it sends the tape directly to the drum for a complete wrap-around, after which the tape encounters two rollers that guide it to the take-up reel.

Though the helical scan opened the way for the development of home VTRs, the first units in which it appeared were anything but inexpensive. Their price tags ranged from $10,000 to $12,000. Over the years, however, improved manufacturing techniques brought the cost down to within the reach of many families. But there is no doubt that the home VTR is still a costly item. It ranges from around $800 up to $2,000 and beyond, with a great many models falling into the $1,200 to $1,500 bracket.

In addition to coming down in price over the years, the home VTR has changed in many ways. It is now more compact than ever before and can fit easily into such confined quarters as school dorms and small apartments. It is

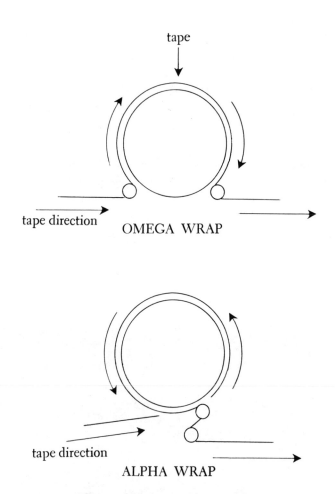

tape

tape direction

OMEGA WRAP

tape direction

ALPHA WRAP

**OMEGA AND ALPHA WRAPS**

now so lightweight that battery-operated, easy-to-carry portables have appeared on the market. It is equipped with features that were difficult, if not impossible, to envision back in 1960. It can even be tied into a small camera so that you can take your own sound pictures. Finally, after using open-reel tape in its earliest versions, it has pretty much switched over to the convenient cassette tape.

Today's cassette tape is usually a half-inch wide and comes in two lengths: T-60 and T-120. As is true of audio cassette lengths, the two designations refer to running times in minutes.

These running times can be adjusted, however, because most up-to-date VTRs are able to operate at three speeds— normal, long, and extended play. At normal play, the tapes run at their posted speeds. Long play doubles the running times to two and four hours respectively. Extended play

**HOME VTR UNIT**
(*Photo courtesy Zenith Radio Corporation*)

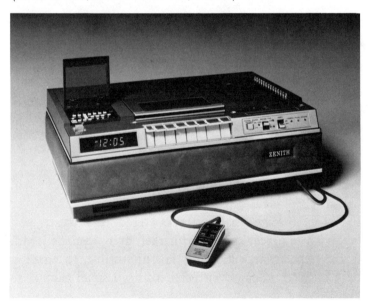

stretches them by yet another one-third, to three hours for the T-60 and six hours for the T-120.

Now what of those machine features that were so difficult to envision back in 1960? A quality VTR today, of course, makes recordings in both black-and-white and color. It is able to record a program on one channel while you're watching another channel. Equipped with a timer that turns the set on and off, it can record programs while you're away from home, automatically changing channels between shows if necessary. Some sets are built to record several shows at different times over the course of a day. Others can record programs over a five- to seven-day period.

Special features may include a high speed—usually a triple speed—playback for searching out a favorite scene in a recorded program. There may also be an audio dubbing device; it enables you to erase the sound portion of a scene, after which you can use an external microphone to put in sound of your own. Some units permit you to transfer home movies to tape for viewing on the TV screen. And there is often a dew detector—circuitry that flashes a warning when the weather is too damp for safe operating conditions and then shuts the unit down to avoid possible damage.

Topping all the features is the camera designed for use with the VTR. Lightweight and compactly built, it can be easily hand-held by means of its pistol grip, or it can be mounted on a tripod. The camera is an optional piece of equipment and is usually purchased separately from the VTR. Its price can easily run to $1,000 or more.

Most manufacturers produce several different cameras in the hope that a customer will be able to find one that suits his or her pocketbook. Depending on cost, the cameras

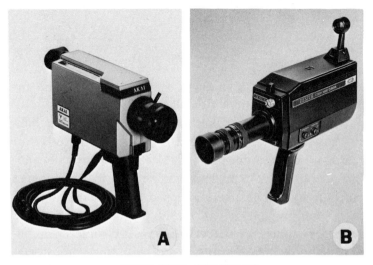

**TWO VTR CAMERAS**

Equipped with pistol grips and capable of taking color pictures, here are two highly popular VTR cameras. The camera in Picture B is outfitted with a zoom lens. Note the two different eyepieces. (*Photo A courtesy Akai America, Ltd.; Photo B courtesy Zenith Radio Corporation*)

may take color pictures or work in black-and-white. All come with a lens that can be adjusted for focus, and all are capable of making a sound recording, with some having built-in microphones while others carry detachable mikes. Higher priced cameras are equipped with a zoom lens—a lens that, by sliding in and out at the touch of a control, enables you to take close-ups and distant shots without moving from your shooting position.

## WHEN YOU WORK THE CAMERA

Thanks to your interest in recording, the chances are good that you'll one day be operating a VTR camera. Perhaps your family already owns one of these little video marvels or is thinking about buying one soon. Or perhaps your school has a VTR unit and is looking for someone to serve as a camera operator for classroom programs.

When the opportunity arises to operate the camera, you'll want to do the best job possible. You'll be able to do so by following a few time-tested suggestions. They'll go on serving you well for years to come.

The first suggestion may seem too obvious for mention, but it must be included because you're likely to be a little on the nervous side. Concentrate always on holding the camera as steady as possible. If you find yourself shaking a bit or starting to hop about in excitement, make a deliberate effort to quiet down. All those shakes and hops are going to result in a recorded picture that trembles.

Next, don't make the classic beginning mistake of thinking that the camera must always be on the move for the picture to be interesting. The basic rule here should be this: Move the camera only when it seems wise or necessary to do so. For instance, settle on the main performer, or performers, of the moment and remain there until the scene shifts elsewhere. Too much camera movement never fails to detract from what is being done or said in a scene.

Take particular care to avoid excess movement if you're working with a zoom lens. There will always be the temptation to overwork the lens because, slipping in so smoothly on the performers, it's great fun to use. But resist the temptation. Remember the basic rule—movement only when wise or necessary.

To get a beginning idea of good camera movement, let's shoot a simple interview at a table. We'll suppose that you're working with a zoom lens and that there are three people at the table—the interviewer and two guests. You can open the scene with an overall shot of the three, staying with it as the interviewer describes the topic to be discussed. Then zoom slowly in on the first guest as he or she is introduced. Let this guest pretty well fill the screen and then pan—a pan is a horizontal movement in either direction—to the second guest for this guest's introduction. At the end of the introduction, zoom back out until all three people are once again in view for the start of the interview.

From now on, the substance of the interview will dictate your moves. If there's some visual material—perhaps a photograph or a piece of equipment—shown at the table, zoom in for a good close-up. Should the interviewer and one of the guests get into a two-way debate, move in a bit on them. If someone rises and moves to another part of the set—say, to a large map—smoothly follow the person.

All these camera moves should be made slowly and deliberately. Never race a zoom or a pan because you're likely to travel too far and then have to suffer the embarrassment of coming back. Further, a fast pan shot will blur the picture, and a zoom that approaches too close can distort a person's features and give the subject a grotesque look.

Now let's talk for a moment about the visual material that's to be shown. Suppose that one of the guests, while seated at the table, is to explain how a small electric motor works. You'll be wise to rehearse with this person for a few minutes before beginning the recording. If the guest is to pick up the motor at any time, urge him or her to hold it in steady hands so that it won't bounce all over the screen and be impossible for the viewers to see clearly; also, the bouncing can make any viewer pretty nervous. And be sure to have the guest keep the motor in one spot; you don't want the thing flying out of the picture at any time. And, if the person is to point to various spots on the motor, ask him or her to use a pencil; then one hand won't accidentally cover up what's to be seen.

Rehearsal is also needed for the guest who must cross to the map. Have this guest move slowly so that you won't need to go racing behind with your pan shot. Coach your subject to stand to one side of the map; the viewers want to see what's there, not what the back of someone's shirt or dress looks like. Should your subject absentmindedly step in front of the map, slowly move to the side so that you can shoot in at a better angle.

If you're working without a zoom lens, you'll need to step forward for close-ups. Unless you take your steps very carefully, the camera will rise and fall with the movement of your body. For a smooth shot, try short, almost shuffling steps while concentrating on holding your torso motionless. Incidentally, rehearse the walk beforehand so that you know exactly how far you'll have to travel. You don't want to bump into the table and fall forward with a live camera in your hands.

One last suggestion: In addition to moving the camera sparingly, don't try to fancy things up with odd camera angles—at least, not until you're an expert cameraman. Simple and honest shots will serve you best. And, even when you are experienced enough for some odd angles, always keep them at a minimum. Strange and off-beat angles may be great for your ego, but, along with too much camera movement, they can distract the viewer and hurt the scene. Your job is to help put the scene across so that everyone can concentrate on what's happening and what's being said. Only when doing your best for the scene—and not for your ego—are you being a truly fine cameraman.

And here we are at last, at the point where there is only one thing left to say in this book: No matter whether there's an audio tape recorder, a deck, or a VTR unit in your future, be sure to have fun—because that's what tape recording is all about.

# SELECTED READING

As YOU develop into an accomplished recordist, you may want to read more about your hobby. You'll find a number of books on recording and stereo equipment at your local library or bookstore. The following suggestions should prove to be especially helpful:

BURSTEIN, HERMAN. *Questions and Answers about Tape Recording.* Blue Ridge Summit, Pennsylvania: Tab Books, 1974.

CLIFFORD, MARTIN. *Microphones: How They Work and How to Use Them.* Blue Ridge Summit, Pennsylvania: Tab Books, 1977.

*Consumer Guide Editors. The Complete Guide to Stereo Equipment.* New York: Simon & Schuster, 1979.

CROWHURST, NORMAN. *ABC's of Tape Recording.* Indianapolis: Sams, 1971.

GAYLORD, M. L. *Hi-Fi for the Enthusiast*. Blue Ridge Summit, Pennsylvania: Tab Books, 1971.

*Institute of High Fidelity. Guide to Stereo High-Fidelity.* Indianapolis: Sams, 1974.

LYTTLE, RICHARD B. *The Complete Beginner's Guide to Hi-Fi.* Garden City, New York: Doubleday, 1981.

SALM, WALTER G. *Tape Recording for Fun and Profit.* Blue Ridge Summit, Pennsylvania: Tab Books, 1969.

SINCLAIR, I. R. *Stereo Cassette Recording.* Rochelle Park, New Jersey: Hayden, 1976.

SWEARER, HARVEY F. *Selecting and Improving Your Hi-Fi System.* Blue Ridge Summit, Pennsylvania: Tab Books, 1974.

WESTCOTT, CHARLES G., AND RICHARD DUBBE. *Tape Recorders: How They Work.* Indianapolis: Sams, 1974.

ZUCKERMAN, ART. *Tape Recording for the Hobbyist.* Indianapolis: Sams, 1977.

# INDEX

123; by home videotape recorder, 179
Playback head, 23
Playback key, 36
Play's the thing, as party game, 115–16
Polyester, 25, 137
Portable cassette recorder, 48, 49; adapter for, 110
Power amplifier in stereo set, 143, 144, 147, 148, 149, 152, 157
Power switch, 34, 49
Pre-amplifier in stereo set, 143, 144, 147, 148, 149, 152
Print-through, 46, 56
Projector: motion-picture, 137, 138; slide, 129, 130, 134–35, 136
Pulse Code Modulation (PCM), 26

Quadraphonic sound, 4-track, 29

Radio: AM/FM, in portable cassette recorder, 49, 50; recording from, 71–74
Receiver, integrated, 147–49
Record changing, manual and automatic, 166, 168, 169
Record key, 35–36, 56, 57, 58, 70
Recording: donor, *see* Donor recording; live, *see* Live recording
Recording gap, 21, 26, 27, 29
Recording head, 20, 21, 23, 32, 36, 53
Recording level controls, 38, 39
Referencing, 37, 38
Remote control, 41
Ribbon driver, 158
Ribbon (velocity) microphone, 81, 83, 104, 105, 106
"Riding the gain," 72, 106

Semiautomatic record changing, 166, 168, 169
Shotgun microphone, 90–91, 110
Shutoff, automatic, 41
Signals, audio, 19, 171
Silent film, adding sound to, 136–39

Singer with guitar, live recording of, 105
Slide projection show, 127–36; background music for, 134, 136; preparation of, 131–35; presentation of, 135–36; synchronization for, 129–30
Solenoids in open-reel unit, 34
Sound stripe on silent film, 136–37, 138, 139
Sound waves, 18–19; electrical waves formed by, 19
Sounds, searching for, 108–11
Splicing kit, 123–24, 125
Splicing machine, 125
Splicing tape, 123
Stereo microphone, 91
Stereo set, 18, 27–29, 142–69; assembly of, 146–56 *passim*; compact, 152, 154–56; components of, 142–46; loudspeakers for, *see* Loudspeakers for stereo set; phonograph in, *see* Phonograph in stereo set; power amplifier in, 143, 144, 147, 148, 149, 152, 157; pre-amplifier in, 143, 144, 147, 148, 149, 152; and tape deck, 15, 143, 144, 150, 152, 154; and three-component system, 152, 153; tuner in, 143, 144, 147, 148, 149, 150, 151, 152
Stereophonic sound, 18, 27–29
Stop key, 36, 51
Stroboscopes of turntable, 163
Stylus for phonograph, 162, 163, 164, 165
Supply reel, 32, 33, 52, 124
Switched questions, as party game, 115
Synchronization: for motion pictures, 137; for slide projection show, 129–30
Synchronizer, 129–30

Take-up reel, 32, 33, 37, 52, 53, 124
Tape: care of, 75–76; digital, 26; editing, 47, 56, 122–26; magnetic, *see* Magnetic tape; storage of, 75–

# ABOUT THE AUTHOR

EDWARD F. DOLAN, JR. has written more than forty books for young people and adults. A fourth-generation Californian, he was born in the San Francisco area and was raised in Los Angeles.

After serving with the 101st Airborne Division in World War II, Mr. Dolan wrote for radio and television. He then worked for a number of years as a newspaper reporter and a magazine editor.

In addition to his books, Mr. Dolan has written numerous magazine articles and stories.

All his books are nonfiction and cover such topics as sports, history, current events, and science and medicine. They include *Starting Soccer, The Complete Beginner's Guide to Gymnastics,* and *Amnesty: The American Puzzle.*

Mr. Dolan and his wife Rose live in northern California. They have two grown children.